Measurement Technology for Process Automation

Measurement Technology for Process Automation

Anders Andersson

CRC Press is an imprint of the
Taylor & Francis Group, an **informa** business

CRC Press
Taylor & Francis Group
6000 Broken Sound Parkway NW, Suite 300
Boca Raton, FL 33487-2742

Printed on acid-free paper

International Standard Book Number-13: 978-1-1380-3539-3 (Paperback)

Library of Congress Cataloging-in-Publication Data

Names: Andersson, Anders (Metrologist), author.
Title: Measurement technology for process automation / Anders Andersson.
Description: Boca Raton : Taylor & Francis, CRC Press, 2017. |
Includes bibliographical references.
Identifiers: LCCN 2017000309| ISBN 9781138035393 (hardback : alk. paper) |
ISBN 9781315267913 (ebook)
Subjects: LCSH: Chemical process control--Automation. | Flow meters. |
Pressure--Measurement. | Temperature measurements.
Classification: LCC TP155.75 .A55 2017 | DDC 660/.2815--dc23
LC record available at https://lccn.loc.gov/2017000309

Visit the Taylor & Francis Web site at
http://www.taylorandfrancis.com

and the CRC Press Web site at
http://www.crcpress.com

Contents

Preface

My idea when writing this book and selecting the layout, was to start at the end. By showing the reader the results first, it may elicit more enthusiasm, and I think being enthusiastic is the best way to achieve good results.

So, in the first chapter you will find examples of what can be achieved by learning things that are described in this book. The text is fairly applied and focuses on practical work with measuring instruments. For more information on fundamental physics and mathematics, there are references to other publications. You can start by finding an example close to what your measuring problem is. In this way, it might not be important in what type of industry you work. For example, measurement and control issues in a mixing process are in most aspects the same, no matter if you work with oil, food or wastewater. After this, you can go on to the specific instrumentation part of the book. If, for example, you need to measure temperature, you will find information about how to select, install and maintain temperature instrumentation in a specific chapter. Then, there are several chapters of general interest for all types of measurements, including sections on electrical signals and safety. The next step is to find out how large your measuring error is, and this is described in the chapters about calibration and measurement uncertainty.

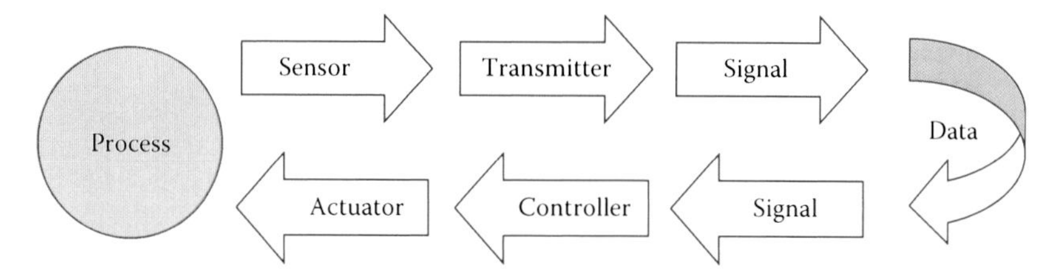

To achieve good measurement data, in short, we can divide the process into four parts:

1. *The instrument itself.* Sensors, hardware and software must be in order and functional. Precision, basic settings and calibration should be adapted to the application with acceptable measuring errors.
2. *Installation and environment.* Install the device correctly. Read the manual and be wary of 'promises' in the marketing material. Good results require that you think about the details! Things like temperature and pressure (both in the process and the environment) can affect the result. So can vibrations, electrical disturbances, electromagnetic radiation and humidity.

3. *Safety*. Check and make sure the instrument will not compromise safety. This involves many aspects, including operation, installation and functionality.
4. *Network*. How is the measuring result sent to the receiver? Irrespective of whether it is a manual reading, an analogue signal or a digital data bus, there are always things to consider and additional uncertainties that (sometimes) can be avoided.

I hope you will find the book useful and that it will help you to produce good and reliable measurement data.

Disclaimer

Every effort has been made to include only accurate and correct information in this book. Nevertheless, no warranties can be made that the text is totally free from errors. Data and conclusions developed by the author are not intended for use without investigation. The author, editor and publisher disclaim all liability or responsibility for damages resulting from the use of information in this book. Readers should always follow legal requirements, local regulations and the manufacturer's advice regarding any installation, operation or maintenance activities.

Texts and examples are primarily expressed in SI units, with references given in other units as well. The comma sign (',') is used as the decimal separator and space will mark thousands (1 000 000).

Acknowledgements

I first started to work with flow meters at Volvo Cars during a school project. It was interesting and challenging. After school I was able to use the project report as reference when applying for a job in the metrology department at SP Technical Research Institute of Sweden. After some years I switched over to a new position at Gustaf Fagerberg AB, a private company supplying instrumentation to the Swedish process industry. All along, I have studied measurement technology at various schools, institutes, companies and industries worldwide. I am very grateful to everyone I met during these years, with each contributing to the knowledge I have today. Thank you all! I would like to express my gratitude to Mrs Kerstin Mattiasson (SP, Sweden), Mr Krister Stolt (SP, Sweden), Dr Masaki Takamoto (Tokyo Keiso, Japan), Dr Friedrich Hofmann (KROHNE, Germany), Mr Bengt Fredriksson (Fagerberg, Sweden), Mr Peter Russel (Evaluation International, UK), Dr Yoshiya Terao (NMIJ, Japan), Mr Per-Ola Olsson (E-ON, Sweden), Dr Jean-Melaine Favenec (EDF, France), Dr John Wright (NIST, USA), Dr Jinn-Haur Shaw (CMS/ITRI, Taiwan) and Prof. Roger Baker (Cambridge University, UK) for their support. Finally, I would like to thank Mr Tim Ennis (Horizon Nuclear Power, UK) for assisting me in writing this book in English, not such an easy task.

Introduction

This book is about process measurement technology. In other words, it discusses activities and equipment that are designed to find out what is happening inside pipes and tanks. To get this information, we need devices like flow meters, pressure sensors, level gauges and converters. If there are exacting requirements on the trueness of the collected information, we need to measure with small uncertainty, and in most processes, true data are equal to good product quality. As most manufacturers strive to achieve higher (or at least constant) quality, most processes require smaller and smaller uncertainties – a development resulting in a need for better measurements! This does not necessarily mean better and more expensive instruments because the result is also very much dependent on other parameters like installation, suitability of selected measuring principle, environment and recalculations. And this is what this book is all about. Learn how to set up a measuring system, what sensors to select, where to put them and how to evaluate the trueness of the obtained result. The book will focus on measurement solutions, installation and electrical signals. If you work with process automation, measurement technology, process design, instrument service or quality assurance, hopefully this book will be of assistance in your everyday work and help you with your problem solving.

Most things described are true worldwide and are in most cases based on international standards. In addition, some local references and resources are listed in Chapter 13.

Anders Andersson
Rydboholm, Sverige

Tim Ennis
English editor

Updates, examples and further information can be found at *www.measurement.academy*.

1

Application Examples

Always remember that safety comes first! But if it is safe to do a test, always try to do a simulation of the measurement or control system you are working on before starting up the real process. The trial-and-error method is probably the best and most efficient way to learn. So, if it is safe, try your ideas in the real world. For this reason, the first chapter in this book is all about practical set-ups, process circuits, control loops and typical installations of meters and sensors.

Hopefully, this will give you both inspiration and some new knowledge!

This book is about process measurement technology. This includes meters for pressure, temperature, level, flow and a few others. The focus is on processes running in pipes and tanks. We will discuss measuring methods, equipment, installation, maintenance and, perhaps most importantly, calibration and traceability.

Safety always comes first! Process pipes and tanks can contain hot, toxic substances at high pressure. Follow safety procedures, use protective equipment and listen to advice and instructions.

Designing a Process

Imagine a factory producing fruit juice. In this process, the people working in the factory use cleaned, rinsed and boiled fruits. The fruit concentrate is then mixed with water and sugar. Finally, the juice is packaged and distributed. This is a rather simple process, which could also be done at home in the kitchen. Everything we do at the factory is the same as what you would do in the kitchen, but now we're going to make juice in greater quantities and in an industrial way. Of course, there are a lot of considerations when building this kind of process, especially when it comes to food and hygiene requirements. But for now, as well as for the rest of this book, we will focus on measurements and the problems around mixing, moving and distributing products, and how to control and measure these processes.

In the example illustrated in Figure 1.1, there are two tanks: one with fruit concentrate (2) and the other with sugar (1). The products are pumped to a new tank (4) where they are mixed with water (3) and heated. When cooking is completed, the bottles are filled with juice (5). To get all this to work, we need four different processes measuring circuits. First, we want to keep track

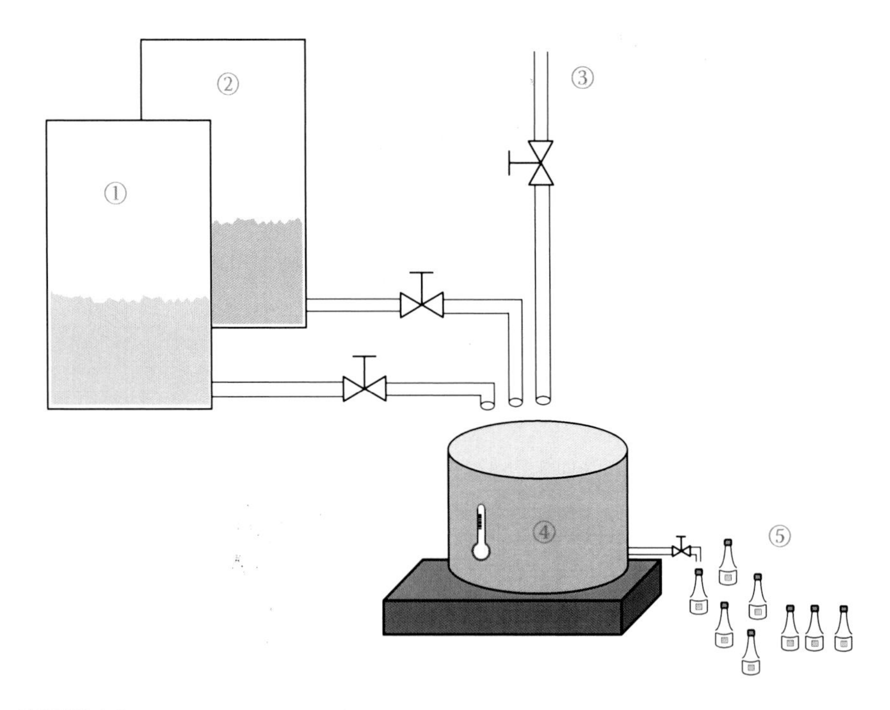

FIGURE 1.1
Mixing and batching.

of how much material is left in tanks (1) and (2). Then, we must mix the ingredients in the right proportions. During heating, we need to have the right temperature. Finally, we would like to know how much we put in each package so that we can put the right price on the products. These are the first four measurement examples on the following pages, where you can read more about how to do this in detail.

In this specific example, it might be that the need for precision is not so important. But basically, the same measuring principles apply to almost any product manufactured. When it comes to costly ingredients, hazardous chemicals, medicines or products with extra high quality requirements, there are likely to be much greater demands on the measurements. Another aspect is that good measurement devices allow you not only to minimise errors but also to verify what has been done with the products, a type of quality traceability that is essential in some processes, like production of pharmaceuticals and food.

Mixing and Batching

Almost all manufacturing processes include mixing of different components. Often a product is mixed with water (diluted) to obtain a final product with the right concentration. However, the purpose of the process can

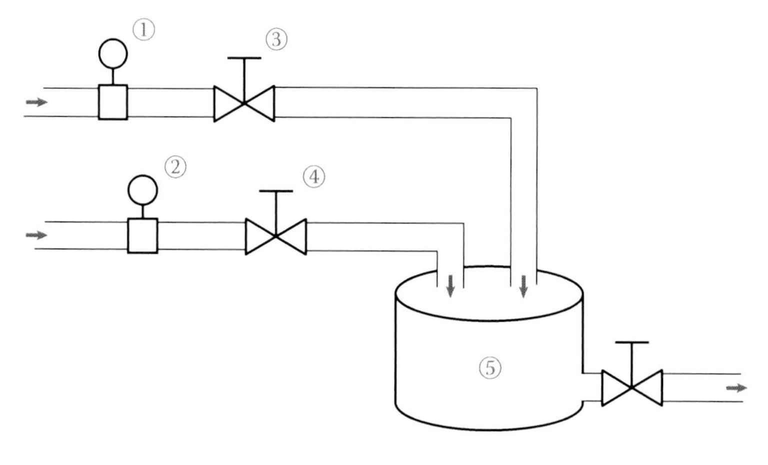

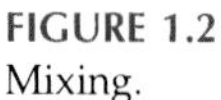

FIGURE 1.2
Mixing.

also be to create a reaction, a particular product characteristic or the right flavour or texture. Process mixing is usually done in two different ways; in batches (in a tank, barrel or similar) or continuously in the 'meeting point' between two pipelines. In many cases, improper dosing of chemicals not only leads to a product of lower quality but can also result in a hazardous product. In such cases, there are of course higher demands on the process measuring equipment in use, to minimise the risk of incorrect mixing and avoid risks.

Two products, A and B, shall be mixed. The mixing is made in a tank (5) (Figure 1.2). As mentioned earlier, this method is similar to the way we would do at home in the kitchen when baking or cookin when ingredients are measured and mixed in a bowl. Here, in this process products A and B can be measured in several different ways, and in the case of large volume measurement, it is probably easiest to open and close valves (3) and (4) in two permanently installed pipelines. To know how long time the valves should be kept open, we can:

1. Read the level change in the mixing tank
2. Use flow (volume) meters (1) and (2) installed in each line

To make the process more automated, the valves can be controlled with preset counters connected to the flow meters. With a preset counter and valve actuators, the only thing the operator needs to do is to press START and the flow will then stop by itself when the right amount is delivered.

Reference, Read More

- To select the best flow meter, see Chapter 2.
- To find an appropriate level meter, see Chapter 5.

- For batching controllers and preset counters, see Chapter 7.
- To select a suitable control valve, see Chapter 8.

Advantage

- Mixing in a tank makes quality control easy because samples can be taken when mixing is ready (and the product is still in the tank).

Warning, Things to Consider

- Mixing in a tank causes interruptions in the production. If the mixing fails, it might happen that the complete batch and all tank content must be destroyed. If the raw materials tend to separate, a stirring device (such as a propeller) might be needed in the tank. Perhaps also cooling or heating is necessary to maintain product quality.
- Select dimensions, response times and filling times that correspond to the requested accuracy. Using level gauges for batch control is often difficult, but sometimes possible if the tanks are high with a small diameter. If using flow meters and high precision is needed, the flow meters must not be installed far away from the mixing point.
- When working with food products, extra care is needed for component and material selection.

Information Needed for Sizing and Selections

- You need to know tank height and diameter, as well as pipe diameter and minimum, normal and maximum flow rates in each line. Media and maximum operating pressure and temperature must be stated. Filling time for each batch might also be of value especially if very short (if less than 1 minute). Before selecting suitable flow meters, an estimate of the required accuracy, based on product quality requirements, is needed. Remember to check material properties when working with chemicals. Select signal input/output types so that sensors and controllers can operate together.

Continuous Mixing

Continuous mixing is a method used in newer production facilities, and often where production capacity is high. In this method, the various ingredients are flowing all the time. Flow rates are measured by flow meters (1) and (2) (Figure 1.3). The mixing is controlled by a control system (5) and two valves (3) and (4) according to a specific recipe, with mixing ratios specified as a %-ratio. The ratio controller can be a part of a large control system or a separate unit. In either case, the additive flow (in our example, the yellow colour) constantly adjusts to follow the main flow (the blue colour) to maintain a steady %-ratio in the final, mixed product (the green colour).

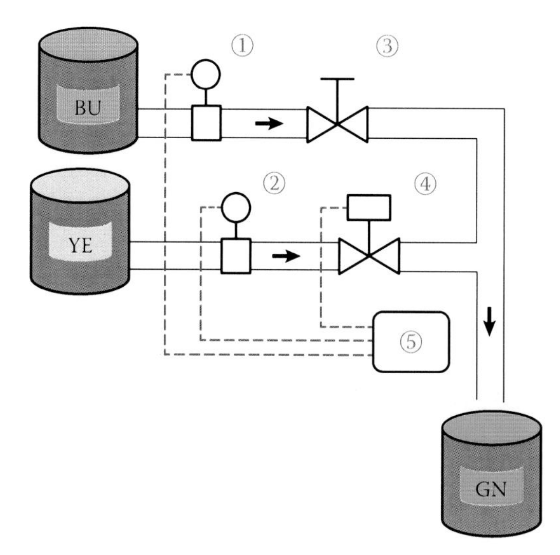

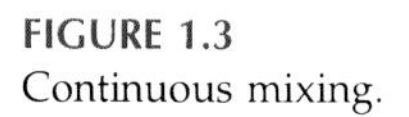
FIGURE 1.3
Continuous mixing.

Reference, Read More

- To select suitable flow meters, see Chapter 2. Basically, all kinds of flow meters can be used, but inductive and mass flow meters are common.
- To select control valves, see Chapter 8.

Advantage

- Continuous mixing allows for higher production, a flexible process with less need of space.

Warning, Things to Consider

- Make sure that products are well mixed. If needed, install a static mixer, a device that starts a rotation in the pipe which helps to mix the product. Product quality can be more difficult to check when the process is constantly running, and therefore a sampling valve is useful.
- Select pipe dimensions, meter sizes and flow rates that correspond to the required accuracy. Avoid long distances between the flow meters and the mixing point.

You need to know tank height and diameter, as well as pipe diameter and minimum, normal and maximum flow rates in each line. Media, maximum operating pressure and temperature must be stated. Before selecting suitable flow meters, an estimate of the required accuracy, based on product quality requirements, is needed. Remember to check material properties when working with chemicals. Select signal input/output types so that sensors and controller can operate together.

Filling

In a filling process, a certain amount (portion) is pumped or transferred into a package (or to another process). The first question to ask when designing such a system is: 'How is each package sold; by weight or by volume?' With constant product quality and known density, it is easy to convert mass to volume (or vice versa), but with varying qualities it becomes more difficult. Therefore, the choice of measurement equipment is more critical when product characteristics vary over time. In the example illustrated in Figure 1.4, we use a circulating line for the product feed. Such a design allows for better temperature and pressure stability in the filling line because the flow in the circulating line never completely shuts down.

These types of processes often run at high speed. In a beverage factory, when filling cans, filling times can be as fast as 1 second. In such cases, it is important that both measuring instruments and valves are fast and matched to each other. As an example, some preset counters will 'learn' (by logging data) how the shutoff valve operates and adjusts the next filling accordingly. It is also important to think about how and where the valve is positioned. The pipeline must be arranged so that the volume in the pipeline between the

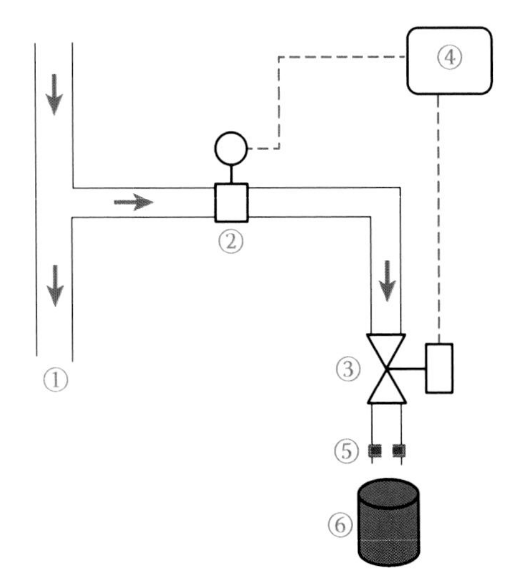

FIGURE 1.4
Filling and batching.

meter and the package does not vary between two fillings. A good way is to have a clearly defined separation (a sort of edge that the liquid can pass) or, in a very fast system, a special valve which closes the pipe end to avoid dripping. For this application, there are also dosing pumps, which are a combination of a flow meter and a pump.

Here, the supply is distributed in a circulating system (1), a sort of overflow where the fluid not used is passed on to other filling lines or back to the storage tank. This is useful even if filling only one package (6) at a time because the circulation will keep the fluid mixed and fresh and it will also maintain a more stable pressure in the supply line. A flow meter (2) will measure fluid passing when the batching valve (3) is open, and a controller (4) connected to the valve and to the output signal from the flow meter will close the valve when the nominal volume is reached. To avoid dripping and getting different volumes between the flow meter and the package (also known as buffer volume), this pipe should be kept as short as possible. It can also be equipped with an 'anti-drip' device (5), a sort of nozzle that will keep the line filled between two packages, when there is no flow.

Reference, Read More

- To select flow meters, see Chapter 2. Inductive, positive displacement and mass flow meters are the most commonly used types.
- To select control valves, see Chapter 8.
- For laws and advice for legal requirements on packed goods, see Chapters 12 and 13.

Advantage

- Continuous mixing allows for higher capacity and a flexible process.

Warning, Things to Consider

- With high-speed filling, it is more difficult to measure with precision! Try to match production capacity to maximum error in weight or volume. In some cases, it is better to arrange two slower filling lines in parallel.
- Select flow meter dimension and flow rate carefully. In a high-speed filling machine, compensate for how the batching valve moves. Install flow meter, valve and outlet nozzle close to each other. Do not use standard settings in electronic meters, instead set time constants, noise filters and similar features according to the specific application.

Information Needed for Sizing and Selections

- Batch size, filling time and fluid are important to know when designing this application. Pressure and temperature are always required when selecting components. Also, try to estimate the required accuracy based on product quality requirements. If products are sold by volume or weight, there may be legal requirements on the application. Select signal input/output types so that sensors and controllers can operate together.

Heat Treatment

When a product is subjected to heat treatment, this is usually done in a furnace or an oven or, if the product is a liquid, in a heated tank. In most cases, the critical aspect of this process is what temperature the product reaches and the time the product is exposed to the heat. In many cases, there are two limits to observe: maximum and minimum. If the temperature is too low, the product treatment may not be good enough. If the temperature gets too high, the product may be damaged or even destroyed. An important technical problem here is that the temperature sensor does not always measure the actual product temperature. Somehow, we instead must use the measured temperature to calculate or estimate the actual product temperature. In the example shown in Figure 1.5, products move on a belt (3) through a furnace. The heater (1) is positioned at the bottom of the oven. By using the average of the four temperature sensors (2), the temperature reading in the controller (4) becomes more representative of the temperature the product is exposed to. However, if the key limit is the highest temperature at which the product is exposed to (perhaps a delicate food product), it is not optimal to control the heat according to the average temperature. In this case, it is better to use a sensor installed somewhere close to the heater (to see the maximum temperature inside the oven) or near the product when it is close to the heater (to see the maximum temperature the product is exposed to).

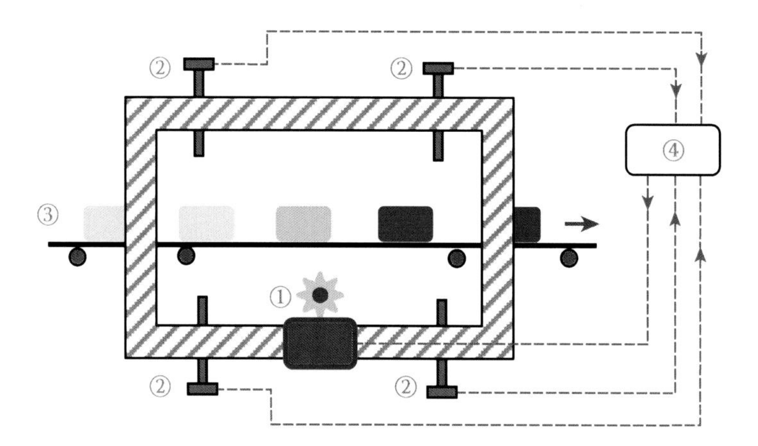

FIGURE 1.5
Furnace/baking oven.

On the contrary, if the purpose is to detect minimum temperature, for example, in a sterilising autoclave, the best position for the temperature sensor is likely to be as far away from the heater as possible.

Reference, Read More

- To select suitable temperature sensors, see Chapter 4. Most common at low temperatures (below 300 °C) are Pt-100-type sensors. For higher temperatures, thermocouples are often used, and in processes at 1000 °C and more (like in the steel industry) contactless, infrared pyrometers (heat cameras) are preferred.

Warning, Things to Consider

- Make sure the product reaches the measured temperature. When testing/verifying the process line, it may be possible to use a moving temperature logger, simulating a product. Remember that all temperature sensors measure their own temperature, not the real product temperature!
- Install sensors in a suitable and representative place, where they will measure a value as close to the required data as possible. Time constants often vary according to installation, sensor design and settings of converters, and all these factors must match the speed of the process.

Information Needed for Sizing and Selections

- When selecting sensors, you need to know temperature ranges and the required accuracy, based on product quality requirements. Select signal input/output types so that sensors and controllers can operate together.

Inventory

Inventory measurements are concerned with knowing how much storage tanks and reservoirs contain. If you know how much material you have on stock, you can set production speed and output accordingly. Just like what you do when you drive your car and plan your route according to how much fuel there is in the tank. The storage of goods in packages can of course be checked by counting bottles and cartons. It is possible that you may have to mix several different measurement principles and working methods to get an overview of the total stock. It is therefore important that one common unit is used to avoid confusion. In most cases, the most suitable unit to use is mass (such as tons) as mass will not vary with temperature and pressure.

In the example illustrated in Figure 1.6, there are two tanks, with the same type of level meter installed. Both tanks hold the same volume, but one is tall

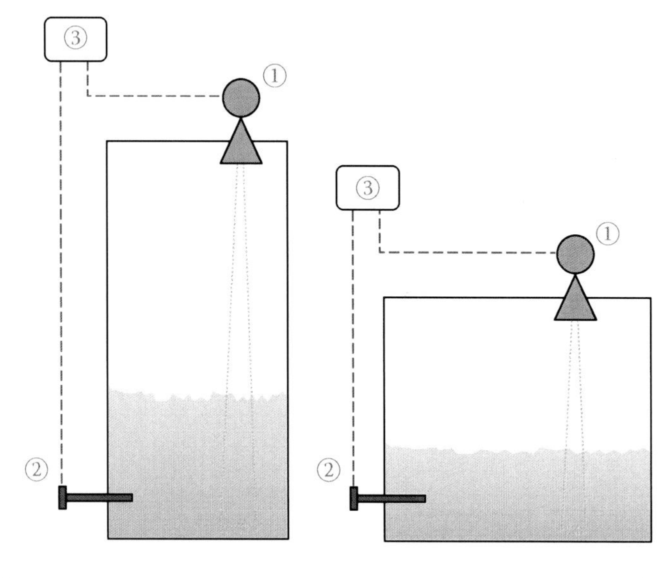

FIGURE 1.6
Level monitoring.

and narrow, the other short and wide. Both level meters have the same specification, but as the accuracy is given in millimetres, the accuracy in terms of volume will be better in the tall tank. This is because the level will change more in this than in the wide tank, for a given volume.

If the temperature in the tank is not constant, the density of the fluid inside the tank will change. And if the density changes, the volume will also change! This means that a slowly changing level measurement does not necessarily mean that there is a leak in the tank; it can be the temperature that is changing. Different liquids have different temperature dependency. Water has a very special density curve, with the maximum density at 4 °C. Petroleum is a product group which has a large density change due to temperature variations. Gasoline, oil and diesel will increase their volume with about 0.2% for every degree Celsius that the temperature rises. Therefore, the level meter (1) in most cases needs to be combined with a temperature sensor (2) and a recalculating unit (3).

Reference, Read More

- To select a good level meter, see Chapter 5. Many different level meters are used in applications like this, where the radar is probably the most accurate solution. Also ultrasonic and pressure sensors are common. To install the complete tank on a weighing platform or on load cells is also possible. Weighing methods and pressure sensors will not detect level changes that are due to density (temperature) variations.
- For overfill protection/safety, see Chapter 9.

Warning, Things to Consider

- If measurements are performed at many points in a large system, for example, to monitor heat balance or total flow of products in a plant, remember to convert all quantities to one single unit. Always select a level meter with the tank shape in mind. Remember that there can be equipment installed inside the tank, such as pipes, stirring devices, ladders and heaters. Such items can cause problems for a top-mounted level meter; the meter will erroneously detect the objects rather than the true level of the liquid. Varying atmosphere inside the tank, with steam, gas or dust, might also result in meter problems and measuring errors for instruments designed for pure air.

Information Needed for Sizing and Selections

- Tank diameter and height are important, as well as operating conditions (media, temperature and pressure). If working with chemicals, check material properties and try to find out how the surface will look inside the tank (calm, with ripples or with foam on top). If you have calibrated tanks, check how to apply the calibration data to the level measurement.

Buying and Selling (Custody Transfer)

Measuring equipment used for selling shall by law (in most countries) be designed according to specific requirements. The purposes of these laws are to prevent fraud and promote fair trade. International recommendations on what methods to use are issued by the International Organization of Legal Metrology (Organisation Internationale de Métrologie Légale, or OIML). Local versions of these documents are then issued in most countries. Most of these different documents state similar basic technical requirements, such as ambient temperature and the ability to withstand external radio signals. Most requirements are also divided in different parts for different meters and/or applications. For example, there are sub-documents regarding domestic water supply, fuel dispensers at petrol stations, weighing devices in grocery stores and taxi meters. The extent of these laws significantly varies from one country to another. In some places, *all* measuring equipment used for payments are subject to legal requirements – even a parking meter. In some other countries, only a few applications are included, for example, volume meters that are used for selling fuel to private persons.

When it comes to legal metrology, liquids such as petrol and diesel belong to a group called 'liquids other than water'. Similar regulations apply for all meters used in such applications. An important fact is that it is not only the meter that fulfils the stated technical demands but also devices like valves, pumps, tanks and pipes around the meter. This is referred to as a 'measuring system'.

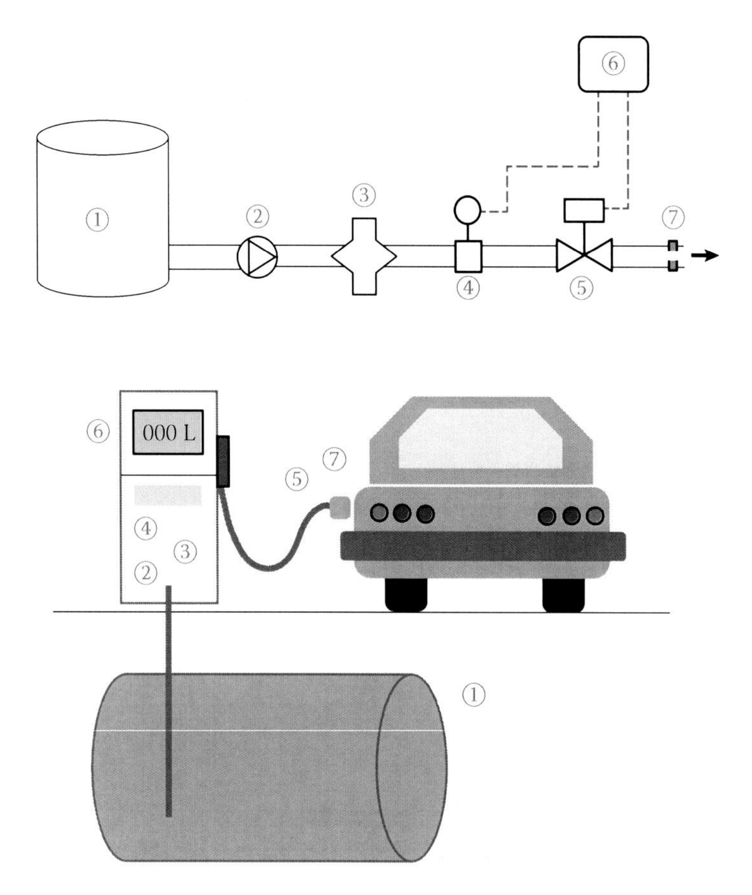

FIGURE 1.7
Custody transfer measuring system.

The example shown in Figure 1.7 illustrates a petrol station. The fuel is stored in underground tanks (1). Each dispenser contains a pump (2) and an air separator or a gas checking device (3). After this, the liquid is measured in a flow (or volume) meter (4). The electronic unit (6) inside the dispenser will control the outlet valve (5) and/or the pump according to input signals from the staff in the shop/kiosk or from a nearby credit card reader. The liquid is pumped via a flexible rubber hose and a nozzle (7) to the car. Inside the nozzle, there are sensors to protect from overfilling and a pressure-controlled valve that will close when the pump is switched off or when the outlet valve is closed. This will keep the hose full and allow the correct volume to be dispensed to every customer.

Reference, Read More

- To select a proper flow meter, see Chapter 2. Many types of flow meters can be used (and are certified), but most common in sales of other liquids than water are positive displacement, ultrasonic and mass flow (Coriolis) meters.

- For laws, directives and regulations that most likely apply, see Chapters 12 and 13.

Advantage

- Selecting a measuring equipment that is type approved (or certified) often results in a safe and secure operation, even if there are no legal requirements.

Warning, Things to Consider

- A certified sensor is not always enough to fulfil legal requirements. In most cases, there is a system approach, and the complete measuring system should be approved. In addition, there is normally a requirement to do re-verification to check measuring errors at specific intervals.

Information Needed for Sizing and Selections

- The type of fluid and nominal flow rate are required when designing this application. Pressure and temperature are always required when selecting components. Most certified measuring instruments are type approved with a system approach, resulting in a need not only for a specific instrument but rather a specific measuring system with various components.

Difference (Net-) Measurement/Leakage Detection

In some applications, it is extremely important that the quantity that enters a pipe (or system) also comes out the other end. For instance, this can apply in a sewage pipe that crosses a lake, a water pipe that is installed under a railway or a pipe with toxic liquid that can cause harm if it leaks. To monitor such installations, two flow meters can be used: one at the inlet and the other at the outlet. If the inlet flow differs from the outlet flow, an alarm signal could be triggered, or a valve at the supply end could be closed automatically. If the pipe is long (or if the system is large), it is almost certain that the two measured flow rates must be compensated for variations in pressure, temperature and other factors that can cause variations in volume between the two measuring points. For such large systems, there is software available, designed to handle such related information, or multi variable inputs.

A similar set-up can be used to measure consumption in an overflow feed system. An example of such a system is the fuel supply to a burner

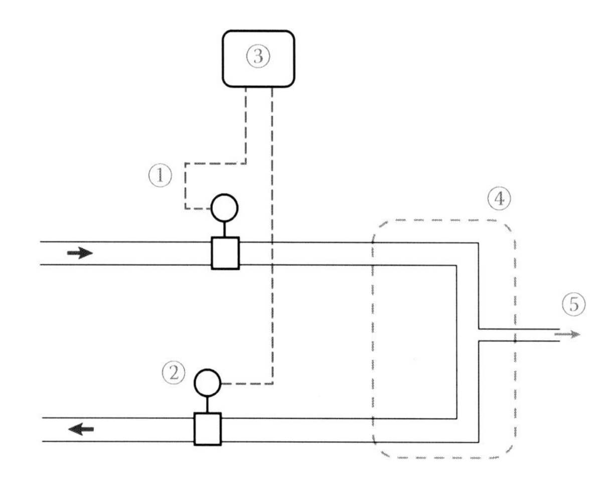

FIGURE 1.8
Net/gross metering.

in a power plant or a ship engine (4) (Figure 1.8). When consumed power is low (at idle), the fuel consumption (5) is low and both inlet flow (1) and return (2) flow are high. At high power, most of the fuel is used and the return flow is low.

In both applications, the dimensioning of the flow meters is very important. As the accuracy of the calculated differential flow depends on the meter reading of both flow meters, the accuracy becomes crucial. Even if the measuring error in each flow meter has a small value in percent, it might result in a large error on the differential flow, such as when the output flow is small. Example: Consumption flow (5) is 100 L/min. Input flow (1) is 1000 L/min and return flow (2) is 900 L/min. The error in each flow meter is stated to be less than 1%. This is equal to 0,01 × 1000 = 10 L/min and 0,01 × 900 = 9 L/min. Basically (without any statistical approach), the error of the consumption indicated in the controller (3) can be up to (10 + 9)/100 = 19%.

Reference, Read More

- To select flow meters, see Chapter 2.

Advantage

- This is a good method to increase safety and reduce the effect of environmental accidents.

Warning, Things to Consider

- Remember that the measuring error in percent might be very high. With a high flow in circulation and a small difference, very accurate flow meters are needed. If pressure and temperature vary between the inlet and the return, it might be required to compensate for those changes.

Information Needed for Sizing and Selections

- The type of fluid and maximum and minimum flow rates for inlet, return and usage are needed when designing this application. Pressure and temperature must always be considered. If working with chemicals, check material properties of wetted components. Check signal types so that sensors and controller will work together.

Air Conditioning/Cooling

Air conditioning systems require lots of energy. In most cases, a cooling system is less efficient than a heating system. Therefore, there is often a lot to gain if a measuring system is installed and used to monitor the efficiency (often stated as coefficient of performance, COP) of a cooling machine or a heat pump. When monitoring the machine, relations between power, speed and temperatures can be observed in real time to optimise the system performance.

In the example illustrated in Figure 1.9, there are several measuring instruments. The primary measurement is the heat meter installed at the inlet (1) using the flow sensor (4D) and the inlet and outlet temperatures (4A) and (4B). In addition, electrical power (3) is measured by an electrical power meter (4C). Produced energy is measured at the output (2) in a similar

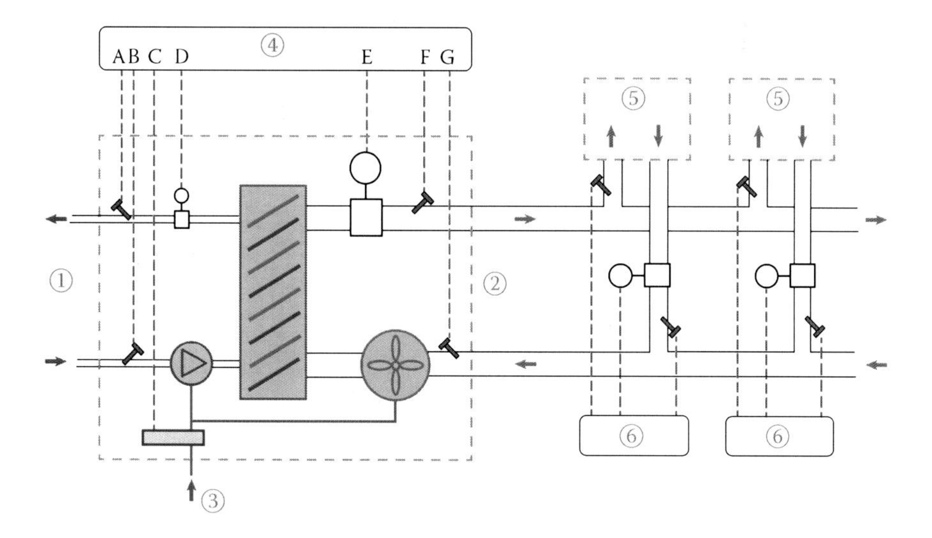

FIGURE 1.9
Heat pump measuring system.

way with (4E), (4F) and (4G). Total production and efficiency can then be calculated and monitored in the central unit (4). Distribution and supply to different consumers (5) are measured by individual heat meters (6).

Reference, Read More

- To select flow meters, see Chapter 2.
- For heat and thermal power calculation, see Chapter 7.
- If used for billing, a heat meter can also be subject to legal requirements – see Chapter 12.

Advantage

- Measuring power and energy in production and distribution networks increases the possibility of detecting energy 'thieves'. Without knowledge and statistics, it is much harder to implement suitable energy efficiency measures. ISO 50001 (Energy Management Systems) describes in general terms suitable actions for energy saving.

Warning, Things to Consider

- When measuring air flow in rooms and large ducts, air velocity can vary a lot depending on the position of the sensor. This is because the flow is not evenly distributed, and your sensor is measuring in one point only. Always try to measure at several positions and calculate an average total flow. Read more about air and gas flow sensors in Chapter 2.

Information Needed for Sizing and Selections

- The type of fluid and the maximum flow rate are required when designing this application. Pressure and temperature are always needed when selecting components. Condensation on cold parts might require special materials (or other procedures for dealing with condensation). Find out if there are legal requirements for the application. Select signal types so that sensors, controllers and instruments will match; also check possible requirements for remote reading.

Pumping

There are several ways to transport a liquid between two places, and pumping is probably the most commonly used method. Different pumps work according to different principles, and the most suitable principle for a particular application is selected based on temperature, pressure, viscosity and whether there are solids or gas bubbles within the medium that is to be transported. Also gases can be pumped (a fan is a type of pump), but others (like steam) are normally not pumped but instead caused to flow by pressure

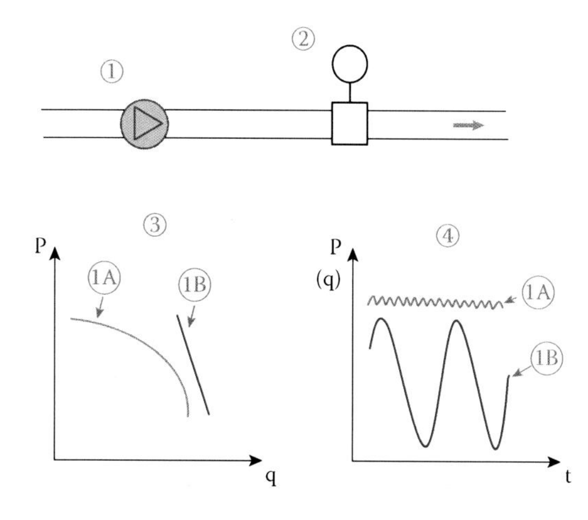

FIGURE 1.10
Pump characteristics.

differences. In applications using pumps, flow meters are often installed in the same line. If the pump and the flow meter are installed close to each other, they can disturb each other, causing production problems or measurement errors. This is especially true when volumetric (positive displacement) type pumps and/or meters are in use.

In Figure 1.10, a pump (1) is feeding a flow meter (2), installed downstream. The relation between pressure and flow rate for two commonly used pump principles is illustrated. The blue line (1A) shows the relation in a centrifugal pump, and the red line (1B) shows the relation in a volumetric (positive displacement) pump. In the first graph (3), you can see the relation between pressure and flow rate (q), and in the second graph (4) pressure and time (t). With a centrifugal pump, there is in general a small, but high frequency, pressure variation in the line. With a volumetric pump, there is a much slower and larger pressure variation. Frequency and amplitude vary with the pump type in use. However, both versions of pump can disturb measuring equipment installed downstream. Downstream of a volumetric pump, it is often necessary to use a specially designed flow meter to give a stable and correct reading.

Reference, Read More

- To select a suitable flow meter, see Chapter 2.

Advantages

- To measure the flow in a pump gives possibilities to monitor efficiency and wear.
- If a small tank with air (or gas with or without a rubber membrane) is installed downstream of the pump, this will allow for compensation

of pressure peaks generated by the pump. Such a device (sometimes called hydrophore) will result in a smoother flow rate, usually good for both measurement and process.

Warning, Things to Consider

- The type of fluid, pressure and temperature will always influence the function of the pump.

Information Needed for Sizing and Selections

- You will need to know pipe diameter and minimum, normal and maximum flow rates. Media, maximum operating pressure and temperature are also needed, as well as frequency of pressure peaks for a volumetric pump. Remember to check material properties when working with chemicals. Select signal input/output types so that sensors and controllers can operate together.

Overfill Protection

If there is too much liquid in a storage tank, all sorts of problems can occur. An overfill protection system is used to prevent poisonous or hazardous liquids leaking from a tank. Therefore, it is in some cases a requirement to install not only a warning system but also an automatic system to shut off valves or pumps. Such a system must of course be reliable and (when overfill protection is a safety requirement) also certified.

Reference, Read More

- To select level meter, see Chapter 5.
- For process safety, see Chapter 9.

Advantage

- Selecting a certified level meter can give you peace of mind, even if there are no legal requirements for it.

Warning, Things to Consider

- To fulfil all requirements, in many cases two independent measuring devices are required. Also, regular service, maintenance and functional tests must be scheduled.

Information Needed for Sizing and Selections

- Tank diameter and height are important, as well as operating conditions (media, temperature and pressure). If working with chemicals, check material properties and try to find out how the surface will

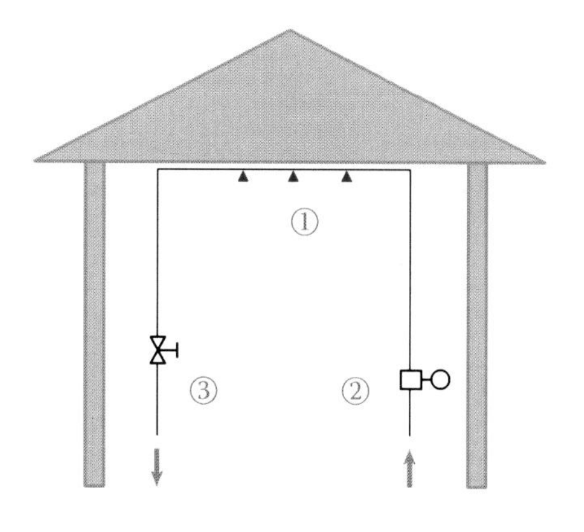

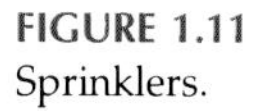
FIGURE 1.11
Sprinklers.

look inside the tank (calm, with ripples, or with foam on top). If needed, do not forget to mention that the application has legal requirements to protect against overfilling.

Sprinklers

Sprinkler systems are used to mitigate the effect of a fire. When the detected temperature in a room reaches a certain limit, an internal valve in the sprinkler system will open and water is released as a spray. Often it is a requirement to test such systems, both the release function and the capacity of the water supply. The release function can be tested by taking samples for testing in a lab. The flow capacity of the water supply to the sprinklers (1) can be tested by opening bypass valves (3) and checking the flow rate measured by locally installed flow meters (2) (Figure 1.11).

Reference, Read More

- To select a suitable flow meter, see Chapter 2. In these type of systems, the flow velocity is very high. This requires well-dimensioned components to avoid large pressure drops. Since the flow meter in this specific application is used to test the maximum water capacity of the sprinkler system, it must have as low a pressure drop as possible. This is to avoid that the meter itself will result in a lower maximum flow (compared to if there had been no flow meter)! Most common types are inductive or ultrasonic flow meters. Also clamp-on ultrasonic meters, installed on the exterior of the pipe, can be used.

Warning, Things to Consider

- In some cases, local government or insurance companies will require regular tests and traceable reference meters. In such applications, a valid calibration certificate must be available for the reference meter. Such a document can be difficult to obtain for a clamp-on meter (even if this in other aspects is a suitable and simple to use device), as in this meter type also the pipework (where the meter is installed) is a part of the measurement.

Information Needed for Sizing and Selections

- State pipe diameter and maximum flow rate as well as maximum pressure. Try to estimate pressure drop and check if it is acceptable.

Pulp and Paper Flow

In principle, a pulp and paper plant is no different from other process industries. However, these plants may have specific measurement applications. One such application is flow measurements of the pulp, a warm liquid containing water, pulp, chemicals and air. The pulp can be difficult to measure and will often require special instruments, or special versions of standard instruments.

Reference, Read More

- To select a suitable flow meter, see Chapter 2. Inductive flow meters are the most common types in this application. With high concentration of fibres and/or chemical additives, electrical noise can occur which requires special actions.

Warning, Things to Consider

- Paper pulp is often at high temperature. That is true also for other process liquids in a paper plant, and because inductive flow meters have liners (internal electrical isolation), it is important to observe limits for temperature and pressure. Especially low pressure, below atmospheric pressure, in combination with high temperature, it can be difficult for the lining material to handle.
- When taking samples, with the purpose of analysing various properties in a laboratory, it is important to take a representative sample. The pulp is not always homogeneously mixed in the pipe; therefore, a special sampling valve can be advantageous.

Information Needed for Sizing and Selections

- The sizing of pipe diameter to normal flow rate is important, both to limit the measuring errors and to keep wear and abrasion low. If working with chemicals, check material properties of wetted components. Additives and particles in the fluid may cause both chemical and electrical noise inside any instrument.

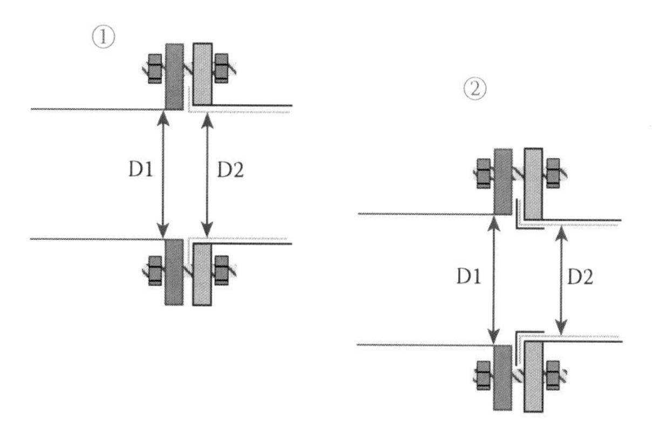

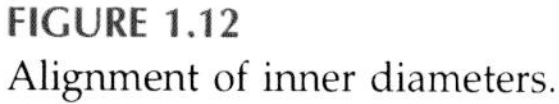

FIGURE 1.12
Alignment of inner diameters.

Slurry Flow

Slurries, pastes and other liquids containing solid particles are frequently used in some process industries, especially in mining. Just as with paper pulp (discussed above), the particles in the slurries and pastes can cause various types of problems. In these applications, clogging, scaling (build-up of internal layers), abrasion and wear must be considered so that the useful life of the instruments will not be shortened. It is important to select the correct size and material to avoid damages to the inner pipe system. Besides the problem of abrasion, the efficiency of the material transport itself is related to velocity. If flow velocity is too high, the water (or transport fluid) will leave solids behind, resulting in clogging. On the contrary, if velocity is too low, particles will fall and pile up at the bottom of the pipe, via deposition, and this will also cause clogging.

In the example shown in Figure 1.12, two different pipe sizes are illustrated. With well-selected dimensions (1), there is no difference in inner diameter when comparing flow meter and pipeline. If the inner diameter of the instrument is smaller compared to the pipeline (2), protection devices are needed. Because of possible measuring errors, generally it is not recommended to use an instrument with a larger inner diameter than the pipe.

Reference, Read More

- To find the best flow meter, see Chapter 2. Inductive flow meters are very common in slurry applications. Solids and particles with specific electrical properties may cause noise on measuring electrodes, and this requires a careful system design. Additionally, flow meters based on Doppler or correlation methods can be used. If the wear can be kept under control, mass flow meters (Coriolis type) can also be used. If so, these can offer extra functionality as they in addition will measure density and in some cases calculate concentration and/or percentage of solids in the product.

Warning, Things to Consider

- Watch out for abrasion and wear. To avoid reducing the lifetime of components too much, it is important to keep a limited velocity in the pipe! Besides the problem of abrasion, also the material transport itself is related to velocity. Optimum velocity depends on material properties; very often around 2 m/s is good enough. It is also important that the inner diameter of the flow meter is the same as the pipe. If the meter has a smaller diameter (if D2 is smaller than D1), metallic protection rings of various designs can be used to improve the situation.

Information Needed for Sizing and Selections

- The sizing of pipe diameter to normal flow rate is important, both to limit measuring errors and to keep wear and abrasion low. If working with chemicals, check material properties of wetted components. Additives and particles in the fluid may cause both chemical and electrical noise inside any instrument.

Cryogenic Flow

A cryogenic application is at a very low temperature. Examples are liquefied gases, such as liquefied nitrogen used in laboratories or liquefied natural gas (LNG) in the energy sector. At ambient pressure, the temperature of LNG is around −162 °C. Because of material properties, pipes, tanks and components will shrink when cooled. If the system is made of different materials, which is quite likely, tensions and forces will occur if they shrink, or contract, at different rates. A flexible section (3) installed somewhere in the line will protect the flow meter (2) and other components (1) (Figure 1.13). The same technique is used when a pipe line sometimes reaches very high temperatures.

Reference, Read More

- For finding a suitable flow meter, see Chapter 2. At such low temperatures, there are high demands on all components, especially the measuring equipment. Installation and mounting positions are even

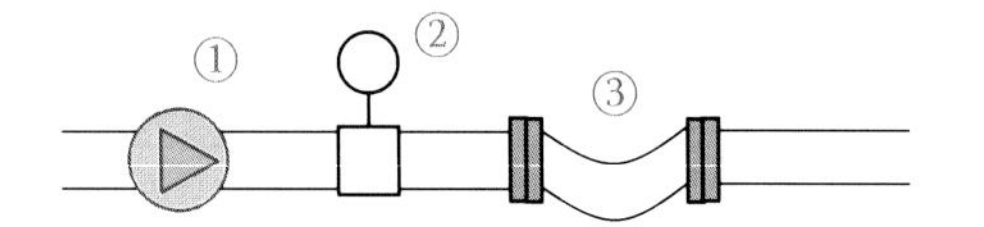

FIGURE 1.13
Flexible installation.

more important than usual to avoid disturbances from the mechanical forces that will occur when materials contract because of decreases in the temperature. Among the most common types are orifice, Coriolis and ultrasonic meters.

- For laws and custody transfer regulations, see Chapters 12 and 13. In some areas, cryogenic systems can have a (clearly visible) bypass valve across the flow meter. This is because the meter and the pipe need to be cooled down so that the fluid is in liquid state before delivery starts (in other CT applications bypass valves are not allowed).

Advantage

- A certified or type approved metering system can offer reliability and confidence, even if there are no legal requirements.

Warning, Things to Consider

- Watch out for heat spots, like badly insulated sections in the pipe system. At such a spot, the liquid can become warmer, and if reaches boiling temperature, gas will form. As the gas is much lower in density, the increase in volume will result in unexpected pressure effects that can push the liquid in any direction.
- An effect like the heat spot described above can be used to protect devices like pressure sensors from the lower temperatures. If installed in a 'dead' T-connection, the pressure sensor will sense the pressure of the cold liquid via warmer gas, without being exposed to cryogenic temperatures. However, in such a case, the drain of the transmitter must never be used, as this would cause cold liquid to enter the transmitter.
- A certified meter is often not good enough. You also need to look at devices around the meter and select suitable components. In addition, regular re-verification is required. Always install instruments so that mechanical forces will not be too large, and use flexible sections if in doubt.

Information Needed for Sizing and Selections

- Low pressure drop is important to avoid gas build-up, and flow range versus pipe diameter must be selected accordingly. As the design temperature of the application is relatively low, the working pressure of the system may be limited. Calibration might be performed using other liquids; check the need to verify this procedure. If legal requirements apply, they are usually very generous, with a large accepted maximum error. Additional requirements on accuracy might therefore be considered as the value of fluid measured might be high.

Sewage Flow

Sewage systems and plants collect and take care of wastewater from houses and industries. Among the special requirements one will find here are the high concentration of solids, large pipe sizes (that are not always completely filled) and long lines where flow is caused by gravity alone (demanding a very low pressure drop). In addition, the accessibility can be low as in most cases both pipes and flow meters are underground.

Reference, Read More

- To select the best flow meter, see Chapter 2. Due to high concentration of solids and the low maximum pressure drop, inductive flow meters are a common choice. There are also ultrasonic Doppler-type meters and flumes and weirs based on equations converting level to flow. In a specially designed flow meter, a level meter can be incorporated to allow measurements even if the pipe is not full.

Warning, Things to Consider

- Local environmental regulations may require certified measuring systems – see Chapters 12 and 13.

Information Needed for Sizing and Selections

- Before selecting a flow meter, specify nominal flow, pipe size and maximum allowed pressure drop. Also specify working conditions and nominal particle content, and check whether regular cleaning is required.

Thermal Power

Thermal power can be distributed by means of flowing hot liquids or steam. District heating is used in many cities where central heat plants supply thermal energy to a group of customers, connected to a piped system filled with process water. District cooling is similar, but here cold water (e.g., from a lake) is distributed. In most cases, a heat exchanger (10) is installed between the piped system (1) and the local system (9) in the building of a customer (Figure 1.14). This allows for water circulation without the mixing of production water and 'user water', and this also offers higher safety in the event of an accident or leaking components. Consumed (or produced) thermal energy can be calculated from flow rate and temperature difference between inlet and outlet. Measurements can be performed at any side of the heat exchanger, but depending on which side of the heat exchanger is used, the results will include or exclude heat losses in the exchanger. The calculation, performed by a

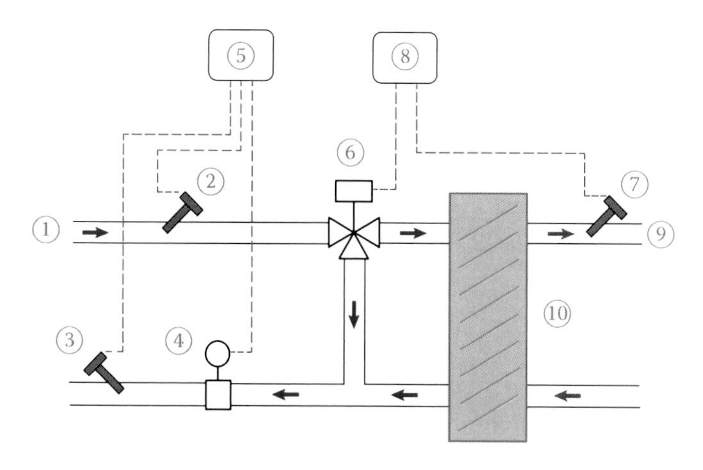

FIGURE 1.14
Heat/thermal power metering.

calculator or computer (5), requires the flow rate (4), inlet temperature (2), outlet temperature (3) and energy content (enthalpy) of the liquid in use. In this example, there is also a controller (8) included. This is connected to the temperature of the outlet on the user side (7) and will try to keep this to a constant level by controlling the flow on the primary side of the heat exchanger, by means of a three-way bypass valve (6). Thermal heat in other applications, like boilers, is measured in a similar way.

Reference, Read More

- To select a suitable flow meter, see Chapter 2. Commonly used in small systems are impeller-type meters (turbines) and in large systems ultrasonic meters. Also inductive meters can be used in systems where the conductivity of the water is not too low and where there is not too much magnetite (magnetite sticking to the inside of an inductive flow meter may cause short circuit of electrodes).
- To select suitable temperature sensors, see Chapter 4. Almost without exception, resistance temperature detectors (RTDs like Pt100/Pt500) are used in these applications.
- For calculation and integration, see Chapter 7.
- For laws and regulations for sales in heating and cooling systems, see Chapters 12 and 13.

Advantage

- Selecting a certified heat meter can provide confidence even if there are no legal requirements. With an output signal from the meter, various control functions can be achieved as well.

Warning, Things to Consider

- To calculate heat from volume flow, the instrument must first recalculate mass flow. This calculation is made based on temperature, and therefore the calculator needs to know the temperature of water when it is in the flow meter. The calculation will require amendment if the liquid contains any impurities or additives, that is, if it is not pure water. So, remember to use another heat calculation equation if the liquid in use is not pure water! Water mixed with glycol or alcohol is often used in cooling systems.
- It is important to always install temperature sensors so that they will measure a representative value, as close to the average liquid temperature as possible. In some cooling systems, the temperature difference between inlet and outlet is extremely low, resulting in a need for very high precision in the temperature measurement.

Information Needed for Sizing and Selections

- Pressure, temperature, pipe diameter and maximum flow rate are needed to select instruments for this application. If the fluid is not water, specify what is used and try to find out whether there is dirt, magnetite or other substances in the pipe work. Select signal types so that sensors and instruments will match, and also check possible requirements for remote reading.

Steam Flow

Steam can be saturated or superheated. Saturated steam is basically what is produced when water boils. Superheated steam has been exposed to more heat and has additional energy. Steam pipes can also often contain liquid water (condensate) after cooling down in long lines or because of production problems. Steam can also contain moisture if it has not been dried and/or separated. For saturated steam, there is a fixed relation between pressure, temperature, density and energy. This means that it is not necessary to measure both pressure and temperature when calculating energy from a volume flow. If the steam is superheated, there is no such given relationship, and here both pressure and temperature are needed for the energy calculation. In Figure 1.15, a heater (1) makes water (2) boil and steam (3) is forming. For safety reasons, the pressure in the boiler tank needs to be monitored and the tank must also be equipped with safety relief valves. During normal operation, the steam will leave the boiler tank and enter the distribution network via a control valve. At all measuring points, volumetric flow, pressure and/or temperature are needed to calculate (4) energy, power and mass.

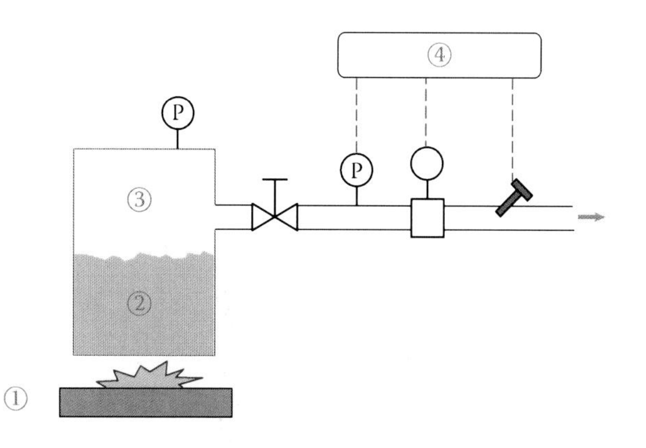

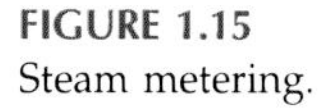
FIGURE 1.15
Steam metering.

Reference, Read More

- To select the optimum flow meter, see Chapter 2. Usual selections include vortex or differential pressure-based flow meters. For superheated steam, ultrasonic flow meters can also be used, if designed for high temperatures.
- To select a calculation device (flow computer), see Chapter 7.

Warning, Things to Consider

- Always select a measuring system including both pressure and temperature sensors if there is any doubt that the steam really is saturated.
- Superheated steam is very hot!. Use components that can withstand heat, or instruments with sensors and electronics that are separated. Never open any ventilation valves. Impulse lines connecting the steam pipe to a pressure sensor will get very hot if steam starts to flow inside. Superheated steam not visible with the naked eye and thus very dangerous to handle.

Information Needed for Sizing and Selections

- Check if the steam is saturated or superheated before selecting an instrument. Maximum and nominal pressure and temperature are always needed when selecting components. Find out what type of data is required: volume flow, mass flow or energy. Select signal types so that sensors and instruments will match.

2

Flow

Basics

When a fluid (gas or liquid) is transported in a pipe, all parts of the fluid may not have the same velocity. The velocity is generally lower at the vicinity of the pipe wall. This variation in velocity is described by the flow profile. It is possible to estimate the flow profile by looking at the shape of the upstream pipe work or by using computational fluid dynamics (CFD) software. If there is optical access (e.g. in free-flowing gas or in pipes with a glass window), the flow profile can also be measured with a laser Doppler meter.

Most flow meters are to some extent dependent on the flow profile. This means that the measuring error will be affected if the flow profile is not as the flow meter designer expected. Some measuring principles are very sensitive, whereas others will be less affected. However, in general, it is important to know the flow profile at the point where a flow meter shall be installed. In a long straight pipe, the flow profile tends to be symmetrical. Figure 2.1 shows two symmetrical flow profiles: one for low flow (1) and the other for high flow (2). Things like bends, T-sections, reducers, filters, valves and sensors will cause unsymmetrical shifts in the flow profile (3). The amount of disturbance or turbulence depends on both pipe material and the fluid itself. There are rules of thumb for various types of flow meters and disturbances. For example, a single bend is usually acceptable if more than five times the pipe diameter is away from an inductive flow meter. The minimum recommended distance is in most cases expressed in number of pipe diameters, for example, 5 × DN (which is equal to 500 mm for a pipe with a nominal diameter of 100 mm). Of course, this minimum distance also depends on how accurate the meter is requested to measure.

To theoretically describe how a fluid is moving, Reynolds number (Re) is often used. Around the year 1900, Osborne Reynolds described the relationship between flow, viscosity and density by an equation and a dimensionless number:

$$Re = \frac{v \times d}{\vartheta} \tag{2.1}$$

where v is the fluid velocity (m/s), d is the diameter (m) and ϑ is the kinematic viscosity (m^2/s).

So, it is not flow rate alone that determines how the flow profile will be, but also density and viscosity are involved. If Re is smaller than around 2000, the flow will be laminar (1). If Re is larger, the flow will be turbulent (2). These two conditions can affect flow meter performance, pressure drop and other parameters.

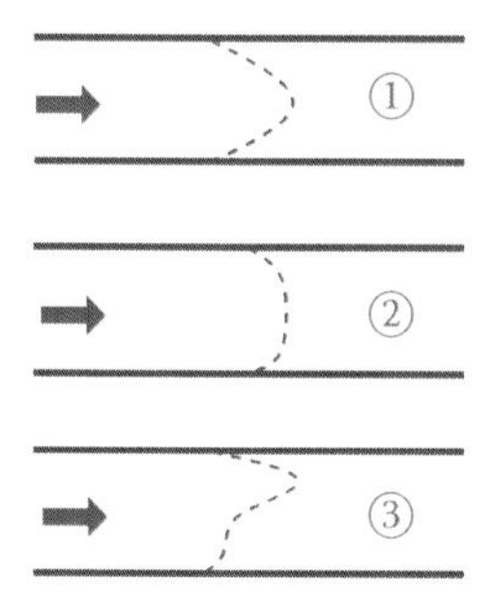

FIGURE 2.1
Flow profiles.

Pressure and flow have a close relationship in a pipeline. This is sometimes used when measuring; flow can be measured to calculate pressure and vice versa. The basics for this were described in the year 1738 by Daniel Bernoulli in his book *Hydrodynamica*. The energy is constant along the pipe (besides what is lost in friction); without leaks, the mass flow rate in a pipe is constant. If, for example, the pipe area is suddenly reduced, the flow velocity will therefore increase to maintain a constant mass flow. To keep the energy constant, the pressure must then decrease at the same time. Bernoulli's equation describes these relations and forms an important base for sizing and design of pipe systems and components. In reality, there are many aspects to consider when describing fluids in motion, but the equation below is the basis for most calculations.

$$P_1 + \frac{\rho_1 \times v_1^2}{2} + \rho_1 \times g \times h_1 = P_2 + \frac{\rho_2 \times v_2^2}{2} + \rho_2 \times g \times h_2 \tag{2.2}$$

where P is the pressure, ρ is the density, v is the fluid velocity, g is the gravity and h is the height.

In Figure 2.2, a pipe is reduced from a large diameter (D_1) to a smaller diameter (D_2). If there are no leaks, the mass flow rate will be constant through the pipe; therefore, the velocity must be higher in the pipe with smaller diameter. In this example, the height is not changing and this part

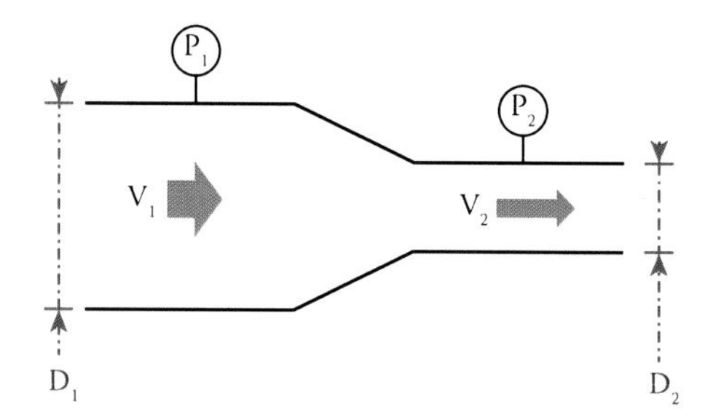

FIGURE 2.2
Pipe area reduction.

of the Bernoulli equation will remain constant. If we look at a flowing liquid, also the density is fairly constant; therefore, the only thing that will change with velocity is pressure. If the velocity increases, the pressure will decrease.

Density and viscosity are two other parameters that will affect flow in pipes. The density of a gas or liquid tells us the weight of a volume unit; for example, water has an approximate density of 1 kg per litre. Viscosity tells us how 'thick' a liquid is, and also how easy it will flow. It is a value of internal friction, and there are many ways to express this. There are two types of viscosities: kinematic and dynamic. The kinematic viscosity is the dynamic viscosity divided by the density:

$$\vartheta = \frac{\mu}{\rho} \tag{2.3}$$

where ϑ is the kinematic viscosity (m^2/s), μ is the dynamic viscosity (Pa s) and ρ is the density (kg/m^3).

The SI unit for kinematic viscosity is m^2/s, and another common unit is centistoke (cSt) (1 cSt = 1 mm^2/s). The kinematic viscosity of water at 20 °C is around 1 cSt. The SI unit for dynamic viscosity is Pa s, and another commonly used unit is centipoise (cP) (1 cP = 1 mPa s). The dynamic viscosity of water at 20 °C is around 1 cP, air has a viscosity of approximately 0,02 cP and olive oil has around 100 cP. You can find the viscosity of selected fluids in Table A3. If temperature increases, the viscosity of a gas will also increase, but in a liquid viscosity will decrease. Gases and most liquids (such as water) have a constant viscosity with respect to velocity, but some liquids are different. If the viscosity changes when a liquid is flowing, this liquid is said to be 'non-Newtonian'. In such liquids the viscosity can both increase and decrease with the flow rate.

Measuring Principles

The ingenuity has been great when it comes to flow meters. There are thousands of patents and ideas related to measuring a flowing liquid or gas, all having advantages and disadvantages. However, the 'perfect' flow meter is yet to be designed! This results in a need to choose the best measuring principle for a specific application, and this makes the flow measurement different and more complicated compared with many other measurement activities. To get a good result, you need to know things such as what media are to be measured, pressure, temperature, pipe size, minimum/normal/maximum flow rate, required accuracy and required response time. Of course, budget and price also need to be included in the selection criteria (Table 2.1).

TABLE 2.1

Flow Meter Selection Chart

Application	Inductive	Ultrasonic	Coriolis	Differential Pressure	Thermal	Vortex	Positive Displacement	Variable Area	Turbine
Water	OK	OK	OK	OK		OK		OK	
Water, condensate		OK	OK	OK		OK		OK	
Water (CT[a])	OK								OK
Thermal heat	OK[b]	OK						OK	
Thermal heat (CT[a])		OK							OK
Petroleum, process		OK	OK	OK			OK	OK	
Petroleum (CT[a])		OK	OK						
Chemical	OK		OK	OK		OK	OK	OK	
Beverages	OK[e]		OK[e]					OK[e]	
High viscosity	OK[f]		OK[e]						
Slurry, paste	OK		OK						
Paper pulp	OK		OK						
Master meter	OK		OK				OK		
Cryo, LNG (CT[a])		OK	OK						
High temperature	OK	OK	OK	OK					
Steam		OK[d]		OK		OK			
Gas/compressed air		OK	OK	OK	OK	OK		OK	

(Continued)

TABLE 2.1 *(CONTINUED)*

Flow Meter Selection Chart

Application	Inductive	Ultrasonic	Coriolis	Differential Pressure	Thermal	Vortex	Positive Displacement	Variable Area	Turbine
Biogas (wet)		OK			OK				
Gas (CT[a])		OK	OK[c]						
Gas x[g] volume		OK							
Gas x[g] mass			OK						

[a] For custody transfer (CT) application, in most cases a type approval is required.
[b] Not suitable if the system contains magnetite/metallic deposits.
[c] Compressed natural gas (CNG).
[d] Superheated.
[e] Special hygienic version.
[f] Based on water.
[g] With varying compositions.

Inductive Flow Meters

Inductive flow meters are one of the most flow measuring commonly used principles in process industries today. The benefits of inductive flow meters are that they have no moving parts (low levels of maintenance required), small measuring error, large measuring range, and low pressure drop and that they will not disturb the flow more than a pipe section with equal length. The sensor is equipped with magnetic coils that produce a magnetic field across the pipe. Inside the pipe, there are measuring electrodes in contact with the liquid and a liner to keep the liquid electrically insulated from the meter body. This meter type will work only on liquids with an electrical conductivity property (oil for example cannot be measured). The meter will primarily measure the velocity of the liquid (in m/s for example), and the flow rate (cubic metres per hour for example) is calculated by an internal signal converter by means of the measured velocity and pipe diameter. Often the flow meter also has a unique factor to compensate for variances in production, like pipe diameter, liner thickness and the strength of the magnetic field. These two values, diameter and meter factor, need to be stored in the signal converter to get a correct measurement.

The signal converter (2) of an inductive flow meter will measure the voltage between the electrodes (1) induced by the magnetic field coils (4) (Figure 2.3). The voltage level depends on the velocity of the liquid, the strength of the magnetic field and the distance between the electrodes (the diameter of the sensor). The conductivity of the liquid is important so that it will match the impedance of the signal converter, but it is not included in the flow equation and a variation will not influence the measurement.

When selecting a suitable inductive flow meter for a specific application, you need to look at the size (diameter) and materials in contact with the liquid ('wetted parts'). Among the available liner materials, you will find rubber, plastic, PTFE (polytetrafluoroethylene) and ceramic. Commonly available electrode materials are stainless steel (for water), various alloys, titanium

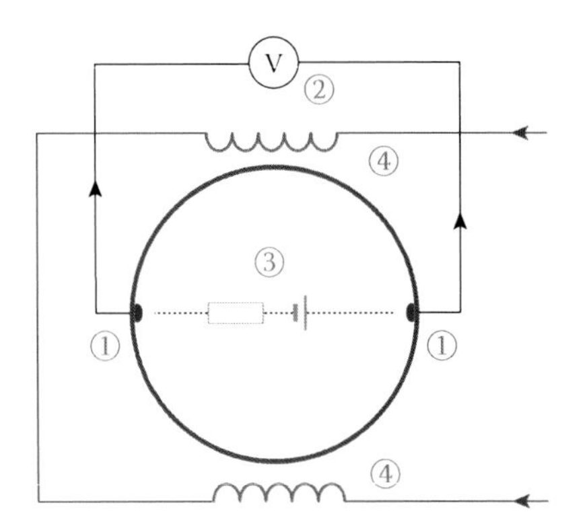

FIGURE 2.3
Inductive flow meter principle.

and tantalum (for corrosive chemicals). Regarding size and dimension, each application needs to be investigated. Do not limit the selection of meter size to pipe dimensions to select meter size, instead try to find out normal operating flow rates from process data. A good target is 3 m/s, even if most meter types will handle velocities up to 10 m/s. With a good inductive flow meter, you should be able to measure with a nominal error less than 0,5% of the measured value. See tables in Appendix for easy conversion between flow and velocity.

As described, the voltage measured on the electrodes inside the meter is proportional to magnetic field strength, distance between the electrodes and liquid velocity. However, a constant magnetic field would cause problems. Therefore, alternating fields are used, constantly changing direction. This results in measuring signals on the electrodes that also change from positive to negative voltages. Early type of meters used AC voltage to generate the magnetic field. Today mainly square wave DC is used, generated and supplied from the internal signal converter. The DC system is in general more precise, and it also consumes less energy. However, a low-power magnetic field will also cause a smaller signal-to-noise ratio, and in some specific cases this can result in disturbances from electrically charged particles in the liquid. To overcome this problem, signal filters and special electrodes can be used.

The voltage supplied to the magnetic coils for the AC system is a sinusoidal wave and for the DC system is a square wave, shown in red (1) (Figure 2.4). The voltage measured on the electrodes (assuming a constant flow rate) is shown in green (2). In both systems, there is a sort of delay time when the direction of the magnetic field changes. In the AC system, this delay has the form of a phase shift (3), and in the DC system there is a sort of 'build-up' time (4). This delay time is not constant with temperature, and it can affect the performance of the flow meter.

Remember that if the liquid you want to measure tends to form crusts or deposits on the interior of the pipe, this will probably affect an inductive flow meter negatively. The magnitude of the error caused by deposits will depend on their electrical properties. Thick isolating deposits will affect all applications and result in unstable flow signals. Conductive deposits, also in thin layers, will cause a negative measuring error. The measuring error

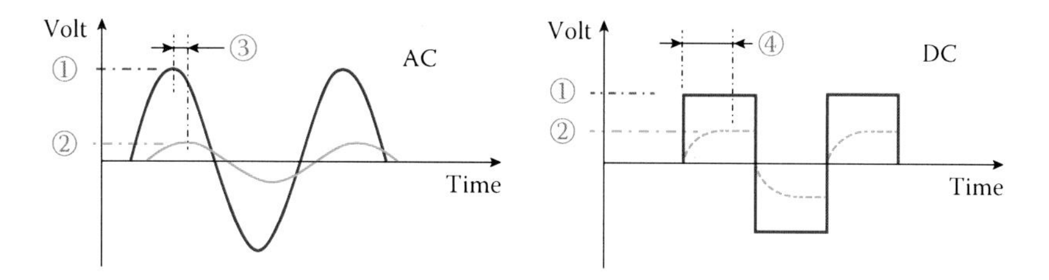

FIGURE 2.4
Induced magnetic field.

is related to the conductivity of the liquid, and a meter measuring a liquid with high electrical conductivity tends to be less affected by conductive deposits.

The voltages measured on the meter electrodes are very small. This requires good measuring circuits, and the signal cable used (between the signal converter and the sensor) must be of good quality. Often several screens and shields are in use to protect the signals from disturbances coming from motors, electrical cables and mobile telephones. The pipeline must be free from electrical voltages (potentials), and a short-circuit cable is in most cases supplied with the meter to protect the sensor from possible voltages in the pipeline. Always make sure that the electrical installation is in accordance with design criteria and installation requirements, which may be described in the flow meter manual.

Ultrasonic Flow Meters

Sound waves can be used in several ways to measure flow. The most common is to use ultrasonic sound and the transit time method. In this method, the time it takes for a pulse to move diagonally across a pipe (1) is measured (2) (Figure 2.5). If the pulse travels both upstream and downstream, it can be measured twice and compared. Differences in time can then be recalculated to flow rate. The average flow velocity along the sound path is automatically measured, but to get the average across the total pipe area several sensor pairs (3) must be used.

Just like inductive flow meters, an ultrasonic transit time meter offers the benefits of no moving parts, low pressure drop, good accuracy and a large measuring range. It will work for all types of liquids, except for those containing many particles (or gas bubbles) as these will stop the sound pulses from crossing the pipe. The transit time ultrasonic meter can also be used to measure gas flow, but not using the same sensors because gas requires sound waves at other frequencies. Sizing and dimensioning are similar to an inductive meter with an aim to get a velocity at about 3 m/s as optimum. When it comes to flow profile and required installation conditions, the need for a

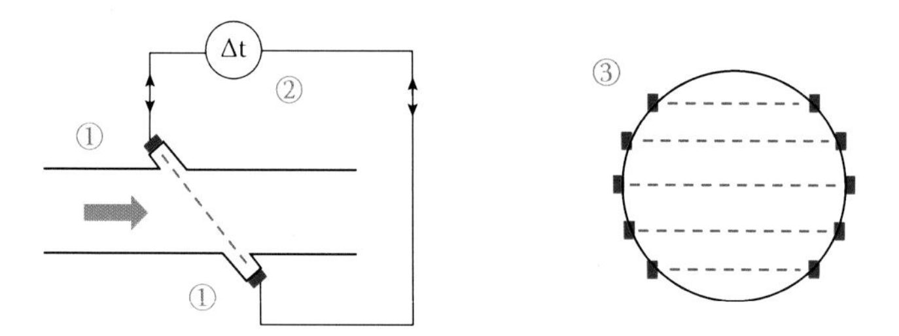

FIGURE 2.5
Ultrasonic flow meter principle.

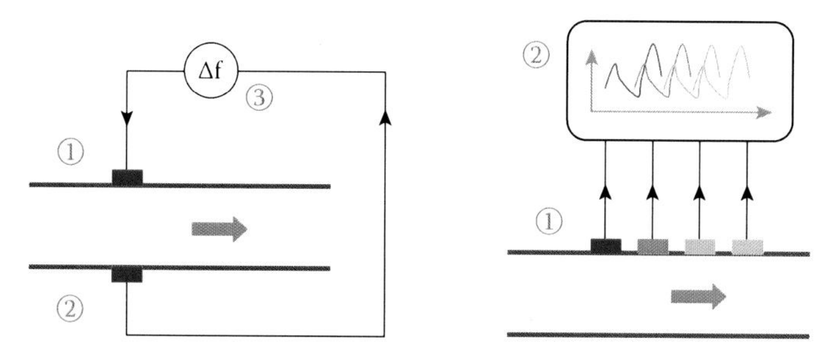

FIGURE 2.6
Doppler and correlation flow meter principles.

straight pipe that runs upstream depends on how many pairs of sensors the meter has. With only one pair (one sound path), the required length is quite large, for example, around 50 × DN after a double bend. A meter with three pairs has similar requirements as an inductive meter, and with 10 or more sensor pairs, there are hardly any requirements at all. Alternatively, if such a multi-path meter is installed in a straight pipe, the metering error can be reduced to a minimum with very low measuring uncertainty as a result. Expected meter errors are in the region of 1% for a single path meter down to less than 0,1% for a multi-path meter. Terminology here is not standardised. Some manufacturers will use the term 'beam' or 'traverse' instead of 'path'. Also 'chord' is sometimes used, perhaps with the meaning that this includes a reflection on the opposite pipe wall. The speed of sound for a few selected materials is presented in Appendix.

Other ways to use ultrasonic sound include the Doppler meter and correlation methods (Figure 2.6). Both are useful in applications using liquids that contain particles or bubbles (or gases with droplets). The Doppler meter measures deviation between transmitted and received frequencies, and this difference is dependent on the velocity of particles in the medium. Necessary presumptions for this method are that the medium contains some particles and that these particles move with the same speed as the medium itself. If there are many large particles, like stones, a correlation meter can be useful. This device will analyse sound patterns at two locations along the pipe. If a pattern from sensor (1) is recognised at sensor (2), the travelling time can be measured and together with the known distance between the sensors a flow rate can be estimated.

Mass Flow Meters

By putting a pipe in vibration, the medium (liquid or gas) inside the pipe is forced to move along with the vibration. The mass and velocity of the media will influence forces affecting the pipe and cause a small deformation. There are various ways to design a Coriolis-type mass flow meter, but the most

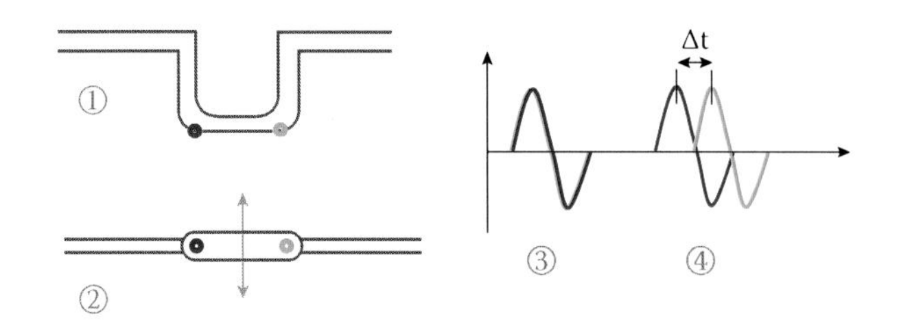

FIGURE 2.7
Coriolis mass flow meter principle.

common ones are made such that both the vibration frequency (2) and the phase shift (4) between the inlet and outlet of a pipe section (1) can be detected (Figure 2.7). As both density and velocity will influence mechanical forces involved, the phase shift will be proportional to the mass flow rate. Due to tensions and misalignments in the pipe system, there might also be a small residual phase shift at zero flow (3), and this is removed by adjusting the zero point after installation. The Coriolis force was first observed on objects in rotation but as the movement of vibrating (or swinging) tubes can be seen as a small part of a rotation, the Coriolis effect will also be seen in a vibration. The pipes can be designed in many ways and you will find meters with different shaped pipes. The trick is to find a good and flexible design that will allow the Coriolis effect and also a strong design that is not affected by external forces and vibrations.

Most Coriolis meters will also measure density. The density of the fluid is calculated from the frequency of the vibrating tubes, as the frequency will depend on the total mass of the oscillating pipe segment. The volume of this segment is known; therefore, the density can be measured.

Even if these two values, mass flow and density, are measured at the same time in the same pipe, they are not totally dependent on each other. The mass flow is a differential measurement and as such is not equally sensitive to external forces and disturbances. The density is an absolute measurement and is therefore generally not very accurate, compared with mass flow reading. As most Coriolis meters can display both mass flow and volume flow, it should be observed that the accuracy of these readings is different; select mass flow if you want the most precise reading.

Even if a Coriolis-type mass flow meter is vibrating, it is considered to have no moving parts. The installation can be somewhat more demanding than other meters when it comes to mechanical support. Especially older meters will require pipe supports attached on a firm and stable base to allow good measuring results. Newer instruments have fewer requirements. There is not any need for a good flow profile in a Coriolis meter, and it can therefore be mounted just downstream of a pipe bend, for example. It can measure all kinds of liquids and gases, and some models will also be able to measure any mix of gas and liquid. However, always try to avoid half-filled pipes as ripple may cause unwanted vibrations. From a good Coriolis flow meter, you should expect measuring errors in the range of 0,2% for mass flow and around 1% for density.

Positive Displacement Flow Meters

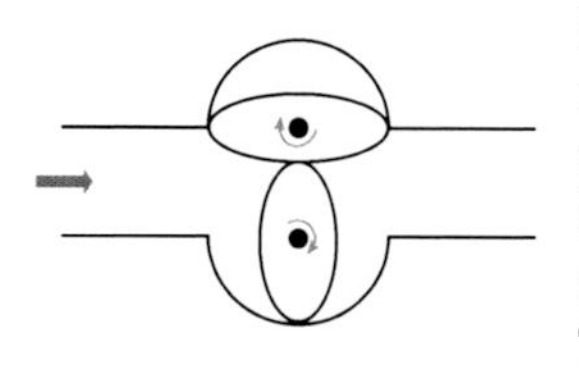

FIGURE 2.8
Positive displacement type flow meter.

Positive displacement meters are sometimes also referred to as volumetric meters, which might provide a better explanation of how they work (Figure 2.8). There are many different designs, but they all have some kind of chamber in common. The chambers are filled and emptied by a mechanical device. The number of times the chambers are filled is counted, and as the volume of the chambers is known, the number can be recalculated to volume and flow rate. If the meter is stopped, the flow will also be stopped. As this is a good thing in an application where liquids are sold, this meter type is very common in fuel dispensers at petrol stations. The meter will cause small pressure variations in the outgoing flow that, in some sensitive processes, can cause problems. This meter type has many moving parts and requires regular maintenance and service. However, if operating in oil (which is quite common), the wear is not significant and the need for maintenance is then reduced. Frequent recalibration is recommended.

Do not use mechanical flow meters in applications where dirt or corrosive conditions can occur, as this will damage sealings and bearings inside the instrument. If needed, install a filter upstream the flow meter. Check that the maximum flow rate in the meter should not be higher than the allowed rate, as excessive flow will damage the meter. Take care during installation, check that the pipework is clean and do not install the meter if flanges and pipes need to be aligned by strong force on the meter.

Turbine Flow Meters

Turbines will rotate with higher speed at higher flow rates. So, the number of rotations is therefore proportional to the volume that passes the meter. However, density and viscosity will also affect the rotation. The sensing element in a turbine meter is often a coil that detects when the blades of the turbine rotate. The output from this sensor is a frequency dependent on the rotational speed. Each turbine is labelled with a 'K-factor' expressed as pulses per litre for a specific fluid. The calibration of a turbine meter is therefore best performed using the same type of media as in which shall be used. This is not easy to achieve, and compensation factors for density and viscosity can sometimes be found. However, the accuracy of these factors is often quite poor. If the meter is well maintained and if it is used and calibrated in the same type

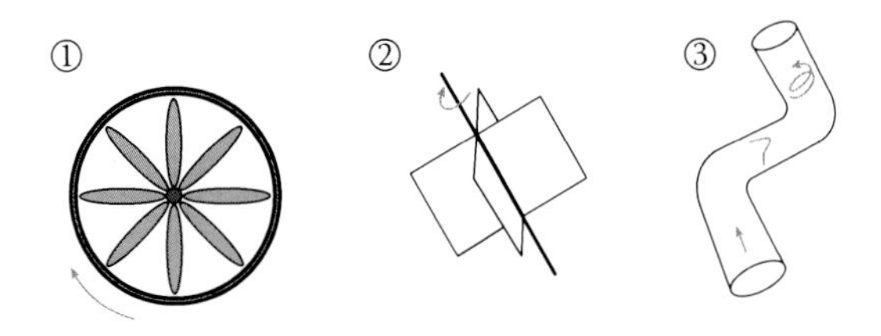

FIGURE 2.9
Turbine, vane and generation of swirl.

of fluid, you can expect a measuring error of around 0.4% and a low response time. The turbine blades are sensitive to particles and bubbles. If particles get stuck on the blades, the speed will increase, resulting in a positive measuring error. The same is true if bubbles are created because of cavitation. To avoid cavitation near the turbine, a back pressure (pressure downstream the meter) is required. The amount of pressure needed varies from meter type to meter type, but pressure in the region of 1 bar g is usually enough.

A turbine meter (1) requires a good flow profile, and it is especially sensitive to rotating flow, for example, caused in a double bend out of plane (3) (Figure 2.9). Quite often a turbine meter needs a flow straightener or conditioner in front to be able to reach its measuring performance.

Very common versions of turbine meters are the ones that we use to measure water consumption in houses. These meters are often designed with a kind of paddle wheel (2), or something that looks like a mix between a paddle wheel and a turbine. These meters are often certified and type approved for use in sales applications where the maximum meter error is 5%.

Differential Pressure Flow Meters

As mentioned previously, Daniel Bernoulli showed us the relation between flow and pressure in a pipeline around 300 years ago. Therefore, measuring pressure loss over a restriction in the pipe (sometimes referred to as a flow element) to calculate flow is quite an old method. In this method, it is easy to design and install the instrument. However, it is not easy to recalculate pressure drop to flow rate. There are several standardised designs for flow restrictions that can be used, and if these are made according to the standard, there are also standardised equations that can be used for the flow calculation. Examples of standardised designs are orifice plates (a disc with a hole) (1) and Venturi tubes (conical tubes) (2) (Figure 2.10). A commonly used standard, which includes both these types, is ISO 5167. When the flow element is designed according to the standard, only a few measures like pipe diameter and hole diameter are needed for the flow calculation. A restriction (like a cone) (3) installed at the centre of the pipe is another method, but since no standardised equations yet exist for this design, calibration is required (the next revision of the ISO standard may also include cone meters).

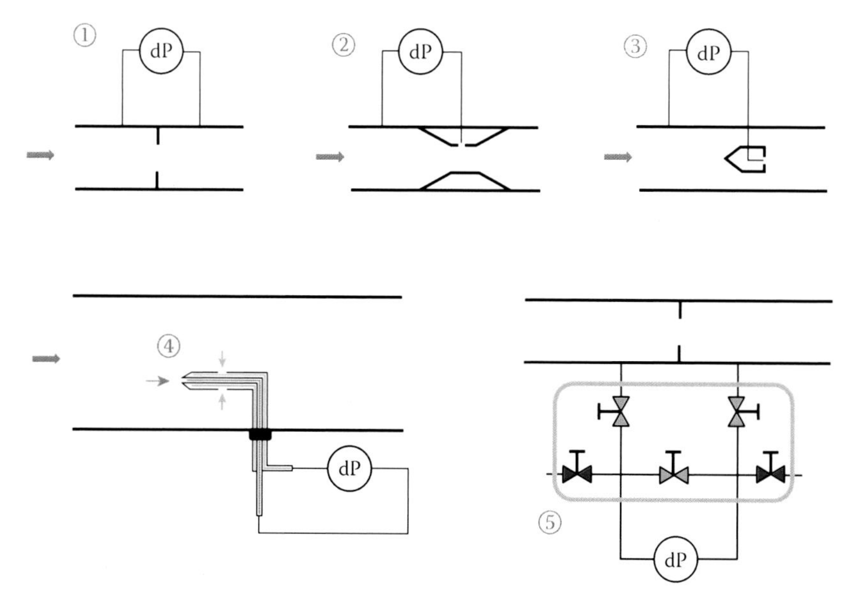

FIGURE 2.10
Differential pressure type flow meters.

The Pitot tube (4) will measure flow velocity at one point only, but by moving the sensor various points can be measured resulting in an average flow. The media will affect the pressure drop. Therefore one needs to know density, viscosity and more to solve the equation. A differential pressure meter (or transmitter) is always needed, no matter the design of the flow element. Installing service valves (valve manifold) (5) between the process and the transmitter is always recommended.

In some new instruments, the flow equation is implemented in the differential pressure meter. This will then automatically calculate a flow rate. As the relation is quadratic, the measuring range is limited (if the flow rate drops to half, the pressure drop is only a quarter compared with full flow). ISO 5167 sets expected measuring errors to the region of 1% for a new flow element without any wear. The error of the differential pressure gauge is additional, but a new electronic dp-transmitter usually has a very small measuring error.

$$v = \sqrt{\frac{2 \times \Delta p}{\rho}} \tag{2.4}$$

where v is the fluid velocity (m/s), Δp is the differential (dynamic) pressure (Pa) and ρ is the density (kg/m^3).

Orifice plates and Venturi tubes require good and undisturbed flow profiles. Long and undisturbed pipelines are therefore required if standard

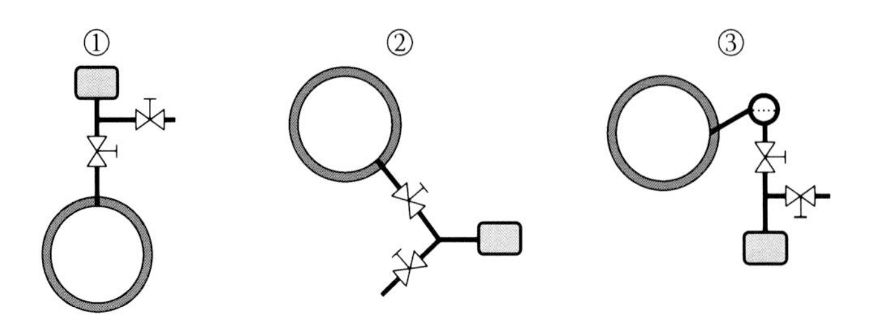

FIGURE 2.11
Pressure tapping and transmitter locations.

equations are to be used. If it is possible to perform a calibration on site, where the relation between flow and pressure is measured, then disturbed flow profiles can be measured. Even existing flow restrictions, like pipe bends, valves or other components, can be used as flow elements, just as long as the relation between differential pressure and flow rate is known.

The position of the pressure transmitter in relation to the flow element can be crucial. Since there are thin pipes (impulse lines) connecting the transmitter to the large pipe (and the flow element), these connections need to be under control. A gas system should not contain any liquid and a liquid system should not contain any gas. Therefore the transmitter should be placed lower than the pipeline when measuring liquid (2) and higher when measuring gas (1) (Figure 2.11). In addition, the pipe should not have any bends that can trap gas/liquid along the way. For steam applications (3), special conditions occur, and depending on pressure and temperature, special 'condensate pots' are recommended to keep the impulse pipes filled with water.

Variable Area Flow Meters

This is a very basic measuring method, based on similar principles used in differential pressure devices, but with a variable area size where the media can flow. It consists of a conical measuring tube and a 'float' (it is called float but it is rather a sinker as it will not float on the media). The meter is mounted in a standing position, where the large diameter of the measuring tube is at the top. As the area where the media can pass the float will increase at higher positions, the float will stay in a position where forces are in equilibrium. This position is affected by density and flow of the media, the weight and shape of the float and the diameter of the tube. As a result, a variable area flow meter (2) will be dependent on media density, and the scale (1) of the meter will therefore specify the application in detail (Figure 2.12). Variable area flow

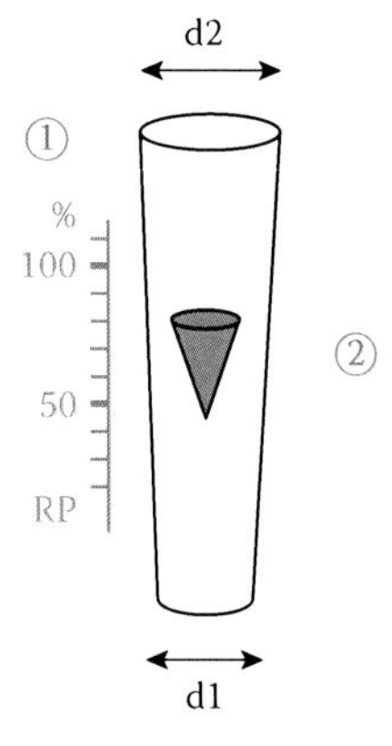

FIGURE 2.12
Variable area type flow meter.

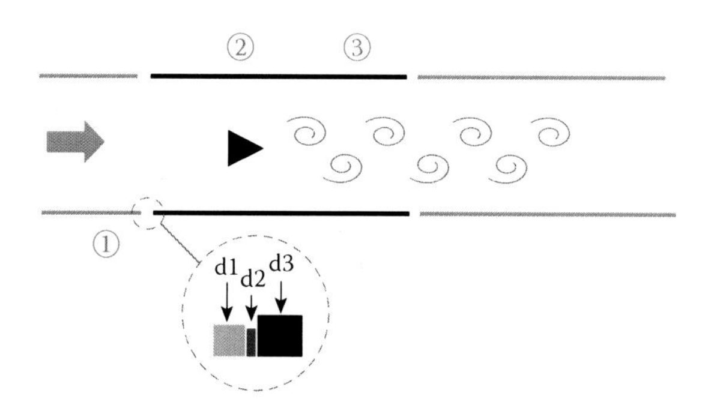

FIGURE 2.13
Inner diameter tolerances.

meters have accuracy in the region of 2%–5%, depending on the length of the measuring tube.

Vortex Flow Meters

A vortex meter utilises one of few principles (maybe the only) that will work with all media types, no matter if it is liquid, gas or steam. Its main application is in steam, as this is a difficult medium to measure for most other flow meters. To create vortices in the streaming flow is rather easy; it is 'just' put a disturbance (called bluff body) inside the pipe. The flow will cause turbulence after this bluff body, and the turbulence will slowly shift from left to right. The frequency of this shift is the vortex frequency, and this is proportional to the velocity of the fluid. At low flow, there is no vortex effect; therefore, there is a minimum required flow under which no measurement is possible (most other flow meter principles have an increasing meter error at low flow). Edges and other parts disturbing the flow near the inlet of the meter can also cause vortexes, and a proper installation is very important. Make sure the inner diameter of the inlet pipe (d1), gasket (d2) and flow meter (d3) matches each other within small tolerances (Figure 2.13).

Thermal Flow Meters

A thermal mass flow meter is often used in ducts and chimneys to measure flow in ventilation systems or smoke going out from a burner. It can be designed in several ways, but a common method is to have two temperature sensors at the tip of the meter probe. One sensor is heated, the other is not. The power required to keep a constant temperature difference is then proportional to mass flow. Another way to do it is to have one sensor exposed to the flow and the other protected. Here, both sensors are heated, and it will take more energy to heat the sensor that is exposed to the flow. Also, in this case the power needed to maintain a constant temperature is proportional to the mass flow rate. This measuring principle has three main advantages: it is relatively good at low flow rates, it is easy to install and the internal design also makes it easier to measure correctly at very small flow rates. Among its disadvantages are that the meter is not very accurate and that there is a need to integrate flow over the pipe area manually as the instrument itself will measure flow velocity in one point only.

General Installation Requirements

As installation is an important issue for almost all flow meters, strongly related to measuring uncertainty, we will repeat some requirements needed for a good installation.

Most flow meter types will improve their performance if they 'see' a good flow profile, if the medium is homogeneous, if pressure and temperature are stable and if there is no mechanical stress in the pipe. Therefore, always check mounting instructions before making the installation work. A nice, straight piece of pipe is always good around a flow meter. The required length of the straight pipe depends on application, measuring principle

TABLE 2.2

Flow Meter Installation Requirements

Measuring Principle	Required Upstream Undisturbed Pipe Length[a]
Inductive	Medium
Inductive, with square cross-section	Short
Ultrasonic, single beam	Long
Ultrasonic, multi-beam	Medium
Coriolis mass flow	Short
Positive displacement	Short
Turbine	Long
Turbine with integrated straightener	Medium
dP: Orifice/Venturi/Pitot	Long
dP: Cone/averaging Pitot	Short
Vortex	Long
Thermal	Long

[a] As indication only. Real requirement very much depends on meter design, flow conditions and pipe configuration upstream of the inlet section.

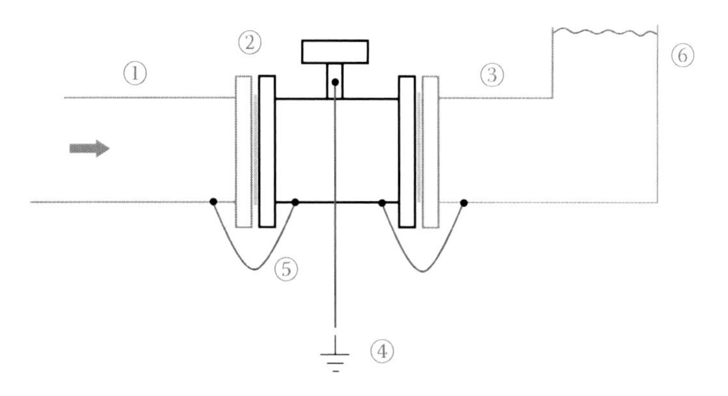

FIGURE 2.14
General flow meter installation requirements.

and required measuring uncertainty but required lengths are probably around those indicated in Table 2.2.

There are a few general installation recommendations valid for all flow meter types. A straight undisturbed pipe (1) before the flow meter is good (Figure 2.14). Also after meter (3), there should be no components installed too close. The flange connection (2) must be free from mechanical forces, the flanges must be parallel, and the gasket in between should be of the correct size (avoid any part protruding in into the flow). The pipeline should be constructed so that the flow meter is always full. Always try to avoid gas in a liquid flow meter and liquid in a gas flow meter. If the meter is supposed to measure batches, a 'separation limit' (6) should be arranged, so that the level (and volume) in the pipe is always the same between batches. Finally, since the flange and the gasket do not always conduct electricity, make sure that the pipes and the flow meter are electrically connected (5) and, if needed, grounded/connected to protective earth (4). If the pipe is already connected to ground, which is normally the case, a separate cable is of course not required.

Calibration and Verification

All flow meter types need to be calibrated! Normally, at the end of the meter production line, the meter manufacturer has a calibration rig where the relation between the flow rate and the output signal is established. Perhaps, orifice plates and Venturi tubes designed and produced according to a standard are an exception as the relation here can be calculated, but the standard still recommends calibration if it is possible. As we have now discussed many times, installation effects can be very important for a flow meter. Therefore, the installation in the calibration rig is of course also very important. Basically, there are two ways to install a flow meter during calibration. The first is to make the installation as good as possible, and this is what most manufacturers do. The other way is to do a simulation of real conditions, a copy (or mock up) of the pipe work where the meter will operate and use this during calibration. As this is a rather time consuming and costly way, this method is used only for applications where requirements are high. Finally, it may be possible to do an in-line, or on-site, calibration. This requires that a flow reference is brought to the site and that this reference can be installed in series with the meter. For small diameter pipes, this is not a very unusual method. Recalibration periods will vary with the application, but intervals of 1 and 2 years are common in the process industry. For custody transfer meters used for sales and/or for tax purposes, there are legal requirements stating maximum reverification intervals.

In any case, the basis for a calibration is a flow (or volume plus time) standard. Depending on the flow meter principle, size and media, the best standard will vary. For some applications (such as very large pipes, high pressure, and high or low temperature), it may be impossible to find a suitable standard, and in these cases, extrapolation of measurements made at

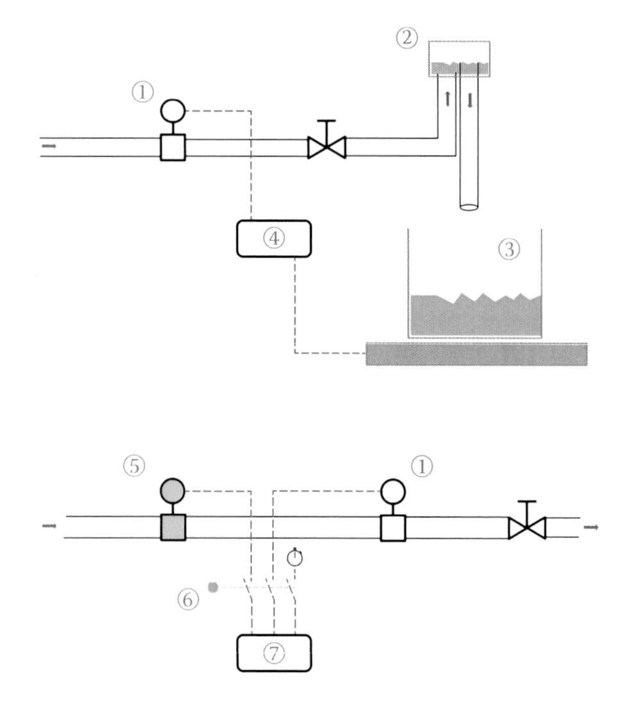

FIGURE 2.15
Weighing tank and master meter calibration.

other conditions must be performed, with an increased uncertainty as a result.

The flow meter to be calibrated (1), sometimes in general aspects called device under test (DUT), is connected to a reference (Figure 2.15). Here, a weighing tank (3) and a master meter (5) are illustrated. In both cases, the reference and the DUT shall be connected in series, and there must be no leaks or shift in temperature in between. The calibration can be 'static' where the flow is stopped before and after each run. Or, it can be 'dynamic' where the flow is running continuously. An overflow device (2) or diverter valve controls the supply to the tank. For a dynamic calibration, the diverter valve also controls the signal recording (4), similar to a master meter system (7). Synchronised recording (6), including time measurement, is required if using a master meter for a dynamic calibration.

Volume Standard

A volume standard is an easy to understand and simple to use device. In detail, it is perhaps somewhat more complicated than it seems, but basically it is just a tank with a known volume. To know that the complete tank is full, there is a need to measure the level. To increase the sensitivity (the relationship between level and volume), it is common to design the volume standard as a bottleneck. By doing so, the level reading will allow for a better accuracy when the volume is close to full. One important limitation in the use of volume tanks is the maximum flow rate. It is simply not possible to fill up a tank in a very short time, at least not under well-defined conditions. As a rule of thumb, minimum filling time is around 1 minute. This is when the flow meter is standing still with zero flow before and after the filling. A way to increase the maximum flow and decreasing the minimum filling time is to use a dynamic measurement method flying start and stop. Here, the flow meter is running with a constant flow before, during and after the measurement. When the measurement starts, the flow is diverted from a return line into the tank by a special valve. When the tank is full, the flow is then again diverted, now back to the return line. The diverter valve is synchronised with an electronic counter so that the flow meter signal is measured at the same time the tank is filling up. The volume scale on a volume standard is in most cases 'wet volume'. This means that it is the volume that can be filled into the tank when the inside walls are wet. Compensation for variation in temperature needs to be done, using standard expansion factor of the tank material. A volume tank with bottle neck normally has an uncertainty of around 0,05%.

A bell prover is another type of volumetric tank, used for gas. Here, the gas is collected under a closed tank, like a bucket put upside down in a bath with water (or oil). The gas will pass the flow meter under test and then go into the tank. As the tank fills up with gas, the liquid is pushed away and the height of the bell will indicate the gas volume.

Weighing Tank

To use a weighing tank instead of a volume tank has some advantages. One is that it is possible to stop the flow anywhere in the tank, without losing too much of the precision. However, density is also required if volume is required. Observe that density inside the flow meter is not always the same as the density inside the weighing tank (e.g. if temperature changes). For a high precision measurement, it is also required to compensate for air buoyancy. You can find more about this in Chapter 5.

Prover

A prover is a semi-automatic machine that measures flow, like a hybrid of a flow meter and a volume tank. In most cases, a prover is made of a pipe (or cylinder) with a known volume. To measure flow, time is required, and a clock is triggered by a moving object. In a ball prover, a rubber ball moves at the same speed as the flow, and the ball will be detected by sensors. In a compact prover, a piston moves with the flow. The piston has a high-resolution position detector that allows its speed to be measured accurately. The advantage of provers is that they can be installed in a closed loop with the flow meter under test mounted in series, minimising the risk of evaporation and loss of liquid in between the test object and the reference. As a prover always uses the flying start and stop method (flow is continuous), maximum flow rate is rather high. Uncertainty is in the same region as a volume tank, below 0,1%.

Master Meters

The easiest way to calibrate a flow meter is to install another flow meter in series and compare measurements from both. However, this master meter must be accurate and reliable and preferably not sensitive to small differences in working conditions. So, the challenge with this method is to find a good flow meter and to be able to verify its function and performance. Special meters are sometimes available, like laminar elements for gas and air. Such devices are made to be used as references and have a good accuracy. A standard process instrument may not be suitable as a master, and to compare two meters of the same kind is not easy. A good example of this is a clamp-on ultrasonic flow meter. This device is very easy to install, as it is mounted on the outside of an existing pipe. However, usually the performance is not as good as an in-line flow meter; therefore, perhaps not very suitable as reference.

Tracers

In the case of very large diameter pipes or when working conditions are harsh, it is not possible to use provers, tanks or master meters. However,

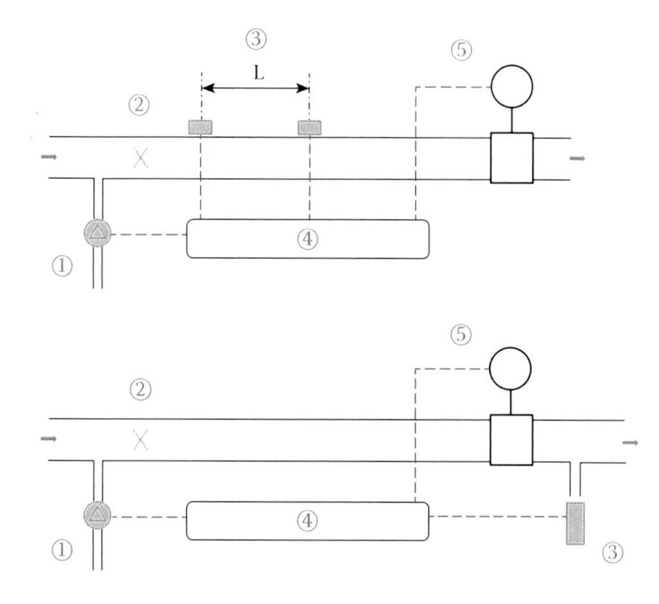

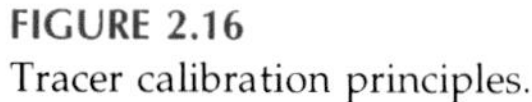
FIGURE 2.16
Tracer calibration principles.

in some of these cases, it can still be possible to use tracers. Tracers are some substances that are normally not present in the process (fluorescence colour, radioactive isotopes or similar). If such a tracer is injected (1) and mixed (2) in the total flow, sensors (3) can detect the speed with which they travel, and by that draw conclusions of the total flow rate (4) (Figure 2.16). If a signal from an existing flow meter (5) is monitored during the same time as the tracer travels through the system, these data can be used as a base for calibration. Normally, an uncertainty of 1–2% can be expected, during good conditions even slightly better. This method is described in ISO Standard 2975.

Further Reading

Roger C. Baker. 2016. *Flow Measurement Handbook*. Cambridge University Press.

3

Pressure

Basics

When molecules in motion collide, for example with the inner wall of a tank, this will cause an impact on the wall. The pressure in this tank is dependent on the strength of these collisions. Higher speed and/or more molecules will result in a higher force on the wall and thus a higher pressure. This is also the reason why the SI unit for pressure is N/m^2, force divided by area. A commonly used unit is Pascal (Pa), and 1 Pa = 1 N/m^2. Another common unit, also accepted in the SI system, is bar, and 1 bar = 100 000 Pa (or 100 kPa). In some countries and industries, psi (pound per square inch) is predominantly used, and 1 psi = 6895 Pa (Table 3.1).

The ambient pressure surrounding us all (atmospheric air pressure) is dependent on the height above the sea and the weather conditions. Even if this pressure varies constantly, scientists have agreed on an average value to use as a reference, normal atmospheric pressure (atm), and this is standardised to 101 325 Pa. At low heights, for example, if you travel up a mountain, the atmospheric pressure will reduce with about 12 Pa/m. If going higher, the atmospheric air pressure is approximately divided by 2 every 5 km. Relatively normal weather conditions will cause pressure variations of around ±5000 Pa. In a precise measurement, for example, a gas metering station, compensation is required for these variations, or design the system so that ambient pressure will not affect the measurement.

Difference due to height is proportional to density, and since water has a much higher density than air, the pressure will change faster in the sea. If you dive, the pressure will increase with around 10 kPa/m.

In a closed tank (without any leakage), the pressure is the same everywhere, and it is not important *where* to measure the pressure as all locations will give the same result. The same is true for a pipe where the flow rate is zero. However, in a pipe where a gas or liquid is flowing, it is different. In this case, there is a pressure drop along the pipe and the pressure will decrease along the pipe. It is possible to calculate the pressure drop, at least approximately, using pipe size, pipe material (surface roughness), flow rate

TABLE 3.1

Pressure Units

Unit	Abbreviation	Value
Pascal	Pa	1 Pa
Bar	bar	100 000 Pa
Hectopascal	hPa	100 Pa
Millibar	mbar	100 Pa
Mercury column	mmHg	133,322 Pa
Water column	mmwc	9,8 Pa[a]
Pound-force per square inch	psi	6895 Pa
Normal atmosphere	atm	101 325 Pa

[a] Depending on reference temperature.

and media properties. Tools and on-line calculators for this are available on the Internet or (more detailed and precise) from professional pipe manufacturers and design/consulting firms. There are several basic equations available, where the Darcy–Weisbach and Colebrook–White equations are most commonly used.

At low Reynolds numbers and at very low pressures near vacuum (below around 100 Pa), other methods and equations are needed.

Pressure Measurement Units

When stating a pressure measurement, there are three things to write down; the measured value, the unit and the base reference (the 'zero level'). There are two common references: vacuum (0 Pa) and normal atmosphere (101 325 Pa).

- Absolute pressure is measured with vacuum as reference, and this is indicated with (a) after the unit.
- Gauge pressure is measured with atmospheric pressure as reference, and this is indicated with (g).

Differential pressure (dP or ΔP) is the pressure difference between two locations, like before and after a pump or at the top and bottom of a tank. This pressure value is not labelled with (a) or (g).

In process automation, mostly gauge pressure is used, and since it is so common, you will not always find the (g) printed. In science and research, mostly absolute pressure is used, and of the same reason also not the (a) is always there. Therefore, it is recommended to be careful so that mistakes are avoided.

Measuring Methods

Unless you work with very high or very low pressures, there are not so many aspects to consider when selecting the best pressure measuring principle for your application. Compared with flow meters, there are just few measuring principles available, and in most process industry applications, the working principle of the sensor is not critical. Installation effects however can still be quite important. At very low pressures, near vacuum, special instruments and installation methods must be used, but these are not explained here.

Gauges

A pressure gauge is often based on a Bourdon tube. Bourdon tubes are basically bent tubes, and when exposed to an internal pressure, they will change in shape. This movement is connected (in most cases with a mechanical link) to a dial where the pressure can be read. Another name for this device is manometer. A manometer does not need external power, it has a low cost and it is robust and easy to use but not very accurate.

Transmitters

A pressure transmitter is a pressure meter consisting of a pressure sensor and an electronic device for amplification and conversion to a standardised signal. They are commonly used in the process industry, and since they produce an electrical output, the main application is process automation. The pressure sensor itself, inside the transmitter, can be of several different principles. Often the sensor is protected with a membrane and not in direct contact with the process. The process media will press on the membrane, and a filling liquid (e.g. silicon oil) behind the membrane will press on the sensor. In some applications, this filling liquid needs to be selected to match the requirements of that specific process.

In Figure 3.1, you can see the housing of a differential pressure transmitter (1) with two connections, high pressure (H) and low pressure (L). On top, there is usually electronics in a separate housing (2), offering standardised output signals and a display. Inside is the pressure sensor (or cell), usually with a membrane. The illustration is excessive, but with high pressure (4), you can see how the membrane is deformed compared with when exposed to low pressure (3). All membranes have two sides. One side is exposed to the process as explained above. The other side can be exposed to ambient pressure (as in a gauge transmitter), to vacuum (as in an absolute transmitter) or to another place in the process (via a second connection, as in a differential pressure transmitter). If there is a second (or outer) membrane in between the process and the pressure cell, the space in between is filled with

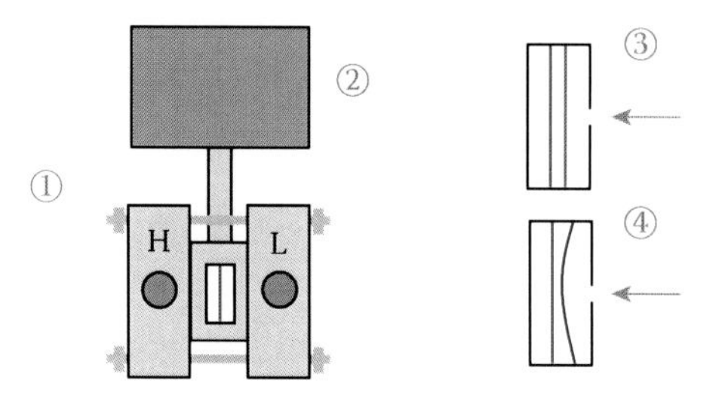

FIGURE 3.1
Pressure transmitter.

substance, in most cases with oil. When the outer membrane is moving because of higher process pressure, the oil is compressed and will push on the sensor. Normally, this is not noticed by the user, but if for example, if the oil expands because of higher ambient temperature or if the transmitter is tilted to another position, the oil itself may cause a pressure indication. The signal will be small, but for a transmitter with a low measuring range, it will be fully detectable. To avoid this, always perform a zero setting with the transmitter in its working position.

There are four common sensor principles available, each one preferred by different manufacturers. A capacitive sensor is based on variation in capacitance when two plates are pushed closer together. The electrical output from this device is almost linear to pressure. Another principle is a strain gauge. This is a small electrical wire that will decrease its diameter if pulled. The decrease in diameter will result in an increase in resistance, which can be measured. Semiconductors are also used, and a piezo resistor works in a similar way. Yet another principle can be compared in a guitar: when force is applied to the string, its frequency will change and therefore measured frequency can be recalculated to pressure.

If your application has a 'nonstandard' requirement (such as response time, long-term stability, ambient temperature and similar), measuring principle is important.

Some pressure transmitters are called multivariable. This refers to the fact that they have an extended ability and can measure more data, for example, differential pressure, pressure and temperature. The purpose is to use all inputs to calculate flow, level or other parameters.

A pressure transmitter can be designed in several ways, but there is a typical model that is used by many manufacturers (see illustration). This common design is made to be mounted near the process and connected to the process via a (or two for differential pressure) tube/small pipe. This pipe is sometimes called impulse line. There are also smaller transmitters (sensors or pressure meters) that are made to be connected directly in the process line.

Installation

Liquid

The transmitter shall be positioned *below* the process. When connected by tubing, the tube shall constantly be directed downwards from the process to the transmitter, so that no gas or air can be trapped inside.

Gas

The transmitter shall be positioned *above* the process. When connected by tubing, the tube shall constantly be directed upwards from the process to the transmitter, so that no liquid can be trapped inside.

Steam

Talking about pressure measurement, steam can be considered in between gas and liquid since the steam at some point will cool down and condense (changing state from gas to liquid). Partly filled tubing in between process and transmitter can cause large measuring errors and must be avoided. One solution to this is to use a 'steam pot' (or condensate chamber), a small tank where condensation will take place and liquid will form at a constant level. The tubing to the transmitter should then *always* be filled with water. If the transmitter is resistant to high temperature, it is also possible to position it above the process, as in a gas application.

Figure 3.2 shows preferred positions for the tapping/pressure sensor connection for gas (1), liquid (2) and steam (3), as well as isolation and vent valves.

Flow Measurement Using a Differential Pressure Transmitter

This is a common application for differential pressure transmitters. To avoid problems from the impulse lines, keep them as short as possible. A valve arrangement between the process and the transmitter is practical for service and maintenance. Diameter of the tubes can be different; smaller diameter will result in a faster response but is also more sensitive to clogging. ISO Standard 2186 recommends using tubes with an inner diameter of 7–10 mm for air, water,

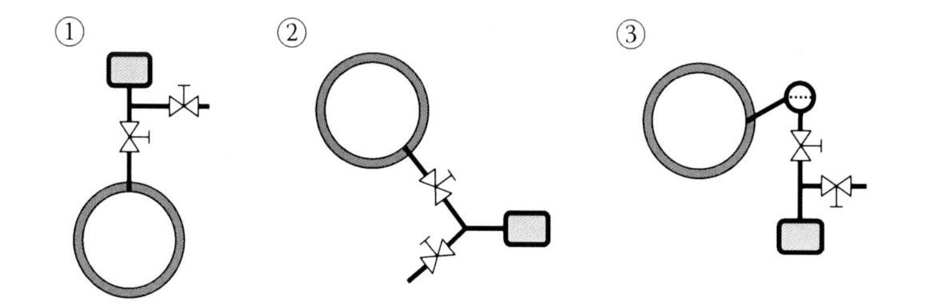

FIGURE 3.2
Pressure tapping and transmitter locations.

steam and dry gas. If distance is long, it is recommended to keep both tubes close to each other, as a difference in temperature might affect the measurement. As an example, 1 m of water column at a temperature changing from 20 °C to 30 °C will affect the pressure with approximately 25 Pa.

Mounting

Make sure that there are no leaks in the tubing between the process and the transmitter. A leakage will result in a flow causing a variation in pressure and measurement errors. An alternative to an impulse line is a remote seal. This is a tube that is pre-filled with liquid, and if installed between the process and the transmitter, it will act as a 'remote control'. It is common to use such remote seals where process media are hot, sticky or abrasive and where the measuring instruments need additional protection from the process media.

The alignment of the transmitter, after mounting, can be important. A pressure transmitter should always be set to zero after all pipes and fastening devices have been tightened. If the transmitter is not standing upright (if there is an angle at any side), this can cause the filling liquid inside the sensor to push unevenly on the membrane and this will affect the measurement. An on-site zero adjustment will compensate for this.

Service Valves

Isolation valves in between the process and the measuring device are always recommended since they allow service and maintenance without the need to shut down the complete process. Single valves can of course be used, one on each tube, but there are also blocks with several valves mounted together and ready to use. Most common are five-valve blocks for differential pressure transmitters and two- or three-valve blocks for ordinary transmitters. Some manufacturers use colour-coded valves, where red means venting (open to atmosphere), blue means isolation (close to process) and green means equalise (connect high and low to each other) (Figure 3.3).

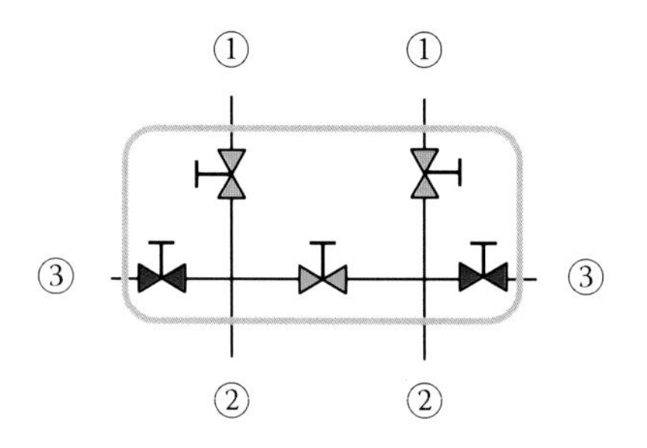

FIGURE 3.3
Service valves for pressure tappings.

Range Setting

A few years ago, all pressure transmitters were delivered without 'absolute calibration'. The signal corresponding to a specific pressure had to be investigated, recorded and set on-site by the user. To do this, a known pressure (normally maximum in that specific process) had to be applied on the sensor, and at the same time, the output signal was set to 100%. Then zero pressure was applied, and the 0% signal was set. Modern transmitters are pre-set and calibrated in the factory before delivery. The sensor is then calibrated at its maximum range, and the user can set any value below this to his or her maximum (100% level). Some manufacturers have kept both methods in their instruments, and it is possible to set the range 'digitally' (by recalculation from factory calibration) or 'manually' (by applying a known pressure).

Calibration and Verification

Probably the most basic calibration method for a pressure meter is a dead weight tester. This device contains a cylinder and a piston loaded by weights. The weights on top of the piston will press the piston downwards, and the force from the total weight divided by the area of the piston is equal to the generated pressure. To be precise, knowledge of the gravitation force (g) in that specific location is also required.

A simpler and more common calibration method is to use a test gauge (3), a calibrated gauge with good accuracy and reliability (Figure 3.4). This is

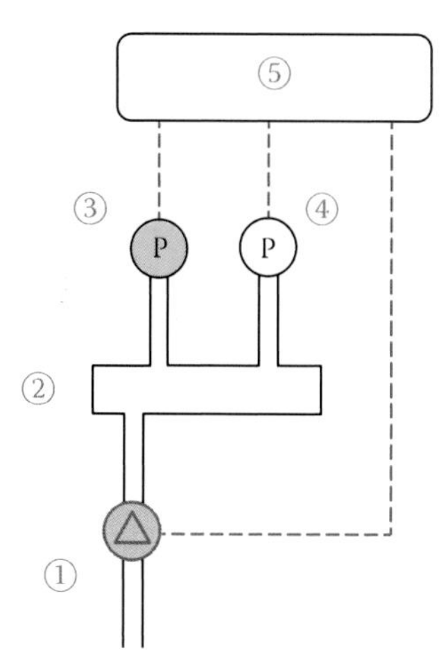

FIGURE 3.4
Pressure meter calibration principle.

connected in parallel, via a manifold (2), to the test object (4), and a pump (1) is used to create the pressure needed for the calibration. The readings of both instruments are then compared (5). To check linearity, it is recommended to compare at a number of pressure levels, cardinal points, between 0% and 100%.

Further Reading

Swedish Meteorological and Hydrological Institute (SMHI). http://www.smhi.se.
National Physical Laboratory (NPL), United Kingdom. http://www.npl.co.uk.
Peter Atkins and Julio de Paula. 2012. *Elements of Physical Chemistry*. Oxford University Press.

4

Temperature

Basics

Temperature is a measure of molecular vibration. More vibration is experienced at a higher temperature, and at the absolute zero point (−273 °C), the molecules are not moving at all. No vibration, no temperature. Several methods, scales and units to measure temperature have been developed, and among the most common are degree Kelvin (based on absolute zero), degree Celsius (based on properties of water) and degree Fahrenheit (partly based on the human body). In Figure 4.1, some commonly known temperatures are shown for these three units: lowest possible temperature, absolute zero (1), freezing point of water (2), human body (3), boiling point of water (4), approximate melting point of steel (5).

Measuring Methods

A few different methods and measuring principles are commonly used to measure temperature. In the process industry, resistance sensors are the most common. At home, the thermistor is the most popular sensor type, used in electronic thermometers. In addition, 'liquid-in-glass' thermometer is the traditional instrument that used to be filled with mercury. There are also contactless, camera-like, infrared sensors. All sensor types can be divided into three groups: mechanical, electrical and optical.

Liquid in Glass

When a liquid is heated other properties like viscosity and density will also change in proportional to the temperature change. If density changes, the volume will also change (mass will remain constant). If a specific mass of liquid is enclosed in a thin tube, the liquid level will increase if the volume increases, and this is how a liquid-in-glass thermometer works. When the temperature changes, the level will change accordingly, and the temperature can be read if an appropriate scale is mounted on the side of the glass.

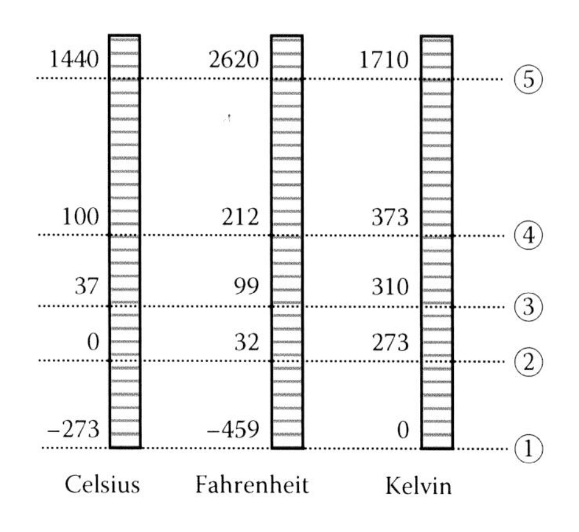

FIGURE 4.1
Temperature scales comparison.

Pt-100

Pt stands for platinum and 100 for a resistance value (100 ohms). A platinum resistance thermometer (PRT) is the most common example where the electrical resistance of the metal is used to measure temperature. Platinum is a very stable material with good electrical properties and is therefore popular in temperature sensors. There are various versions of the PRT such as Pt-25 and Pt-1000. The variation in nominal resistance does not tell us about the quality of the sensor, for which there are different tolerance classes. The platinum can be attached to a circuit board or in the form of a thin wire, either hanging loose or winded on a spool inside the sensor. The resistance of a Pt-100 sensor will change approximately 0.4 ohm per degree C. A more precise value of the resistance (R) can be calculated by the following equation:

$$R = R_0(1 + 3{,}9083 \cdot 10^{-3} \times t - 5{,}775 \cdot 10^{-7} \times t^2) \quad (4.1)$$

where t is the temperature in degree C and R_0 the resistance of the sensor at 0 °C.

Electrical Connection

The resistance of a Pt-100 sensor is measured by supplying a small current (typically in the order of 1 mA) through the sensor, and by measuring the voltage drop, the resistance can be calculated. As instrument (3) that measures voltage is often at some distance from the sensor (1), connecting cables (2) will also influence the result (Figure 4.2). If two wires are used, the cable resistance will simply be added to the temperature. When using a coarse copper cable of 0,75 mm^2, the measurement error is around 0,1 °C per metre.

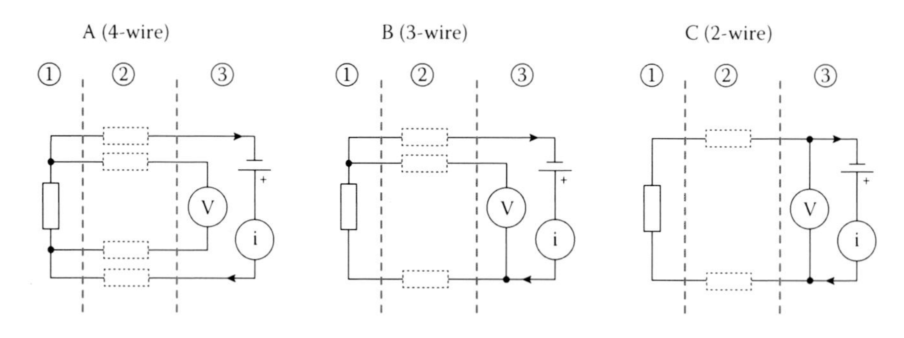

FIGURE 4.2
Resistance measurement circuits.

To avoid this additional temperature reading (or measurement error), three or four wires can be used. The measurement current is now separated from the volt meter; therefore, cable resistance is kept separate. For accurate measurements, four-wire connection (and instrumentation) is strongly recommended.

Thermocouple

A thermocouple is a temperature sensor consisting of two wires made of different materials. The wires are connected at the tip of the sensor, a junction, and in the other end there is a volt meter. When this sensor (or thermocouple cable) is exposed to temperature variation along its length, a voltage proportional to the temperature is generated. As the 'starting temperature' is the ambient temperature, this temperature must be added to the measured value. This is often called 'cold junction compensation', and a feature that can handle this is included in most commercial temperature transmitters. Compared with the resistance sensor, one difference with a thermocouple is that the signal will be generated along the entire cable, at all sections where there is a temperature change, not 'only' in the sensor tip as in a resistance sensor. Many different materials can be used, and the sensitivity to temperature and environmental conditions needs to be considered when selecting a suitable type (pair of materials).

Infrared Thermometers

A convenient and contactless method to measure temperature is to look at heat radiation. All objects will emit infrared energy, and with increasing temperature, the radiation will also increase. An infrared pyrometer can collect the radiated energy and recalculate this to temperature. However, the collected energy is dependent not only on temperature but also on emissivity and transmission. The emissivity is a measure of surface characteristic, ranging from 0 to 1. A 'shiny' metallic surface has an emissivity around 0,1 and a black dull surface has around 0,9. The transmission depends on what substances there are in the atmosphere in between the object to measure and the pyrometer. Steam, smoke and wet gas will reduce the transmission, and the damping effect depends on both concentration and the wavelength of the infrared energy the sensor is looking at. To be able to calculate temperature from energy radiation, the surface of the measured object must be known, as well as the atmosphere in front of the sensor. If the sensor can look at several wavelengths, it might be possible to automatically detect, and compensate for, variations in emissivity and transmission. The emissivity of a few materials is listed in Appendix.

Installation

A temperature sensor will measure the immediate surrounding temperature only! Most likely, this is not what is required; you will want information about the temperature of an object, a liquid or a process. Good thermal connection (heat exchange) between the sensor and the process must therefore be ensured. There is never any sharp 'border' between different temperatures; there is always a zone where gradients will form. With good insulation materials, the thickness of this zone, in an oven for example, can be rather small but not zero. In a similar way, the temperature sensor itself will form a gradient zone with a temperature varying from process level near the tip of the sensor to ambient temperature at the connection end. If the process is warmer than the surrounding temperature, the measured value will be too low because heat is transported away via the sensor. To avoid or minimise this problem, longer sensors and sensors with a better thermal contact can be used. Insulation of the part of the sensor that is outside the pipe might also be effective.

Pipe Mounting

When the temperature sensor is installed in a pipe, the sensor tip should be exposed to the flow, with as little 'thermal disturbance' from the environment as possible. There are a few but effective basic recommendations on how to mount the sensor (see Figure 4.3). If the flowing liquid or gas has a non-homogeneous temperature distribution, it can be difficult to find the best position; in these cases, several sensors might be needed to get a good average representative temperature.

Three different mounting positions can be seen to the right: excellent (A), very good (B) and good (C) mounting positions of a temperature sensor in a pipeline. To the left, you can see the tip of the sensor, with the sensor element (1) and electrical cables (5) in the middle of a protection tube (4) inside a thermowell/outer protection tube (3). Some sensors are fitted with a socket (2) to improve the internal thermal conduction.

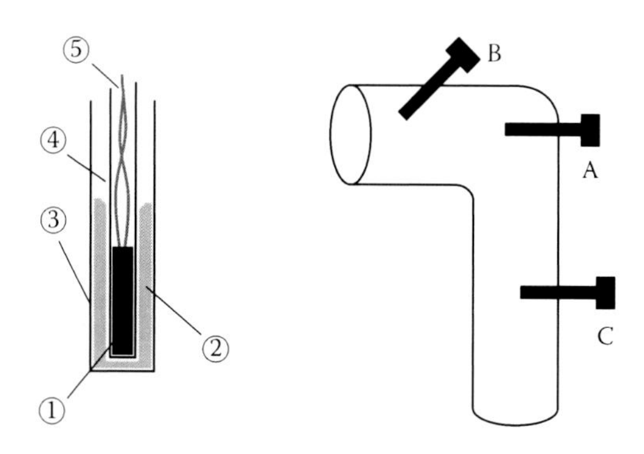

FIGURE 4.3
Temperature sensor design and pipe mounting positions.

Thermal conductivity in general plays an important role for temperature measurement, not only inside the sensor but also outside the sensor. The sensor will measure its internal temperature! With a good thermal conductivity around the sensor, this internal temperature is likely to be the same as the surrounding temperature, which is in most cases what is desired. With low thermal conductivity, there will be a difference, resulting in an increased measurement uncertainty.

Calibration and Verification

As mentioned, the base for the Celsius temperature scale is the freezing and boiling temperature of water. In the International Temperature Scale (ITS), freezing, melting and boiling points of other materials are listed. These points are used as references in temperature measurement laboratories. To create a foundation for traceable temperature calibrations, very pure substances that can be melted and measured are therefore needed. Ambient pressure will affect the boiling temperature, but not the freezing point. It seems quite strange, but this means that if pressure is low enough, a liquid can boil and freeze at the same temperature. This condition is called the triple point, where, for example, water is in solid (ice), liquid and gas (steam) phase at the same time. The materials listed in ITS tables cover an approximate range from −270 °C to 1500 °C. Reference temperatures according to ITS are listed in Appendix.

Comparison

When performing a temperature sensor calibration, the idea is to compare the temperature sensor you want to test with a 'better' and 'more reliable' sensor: a master! As all calibrations need to be traceable, there is a need to

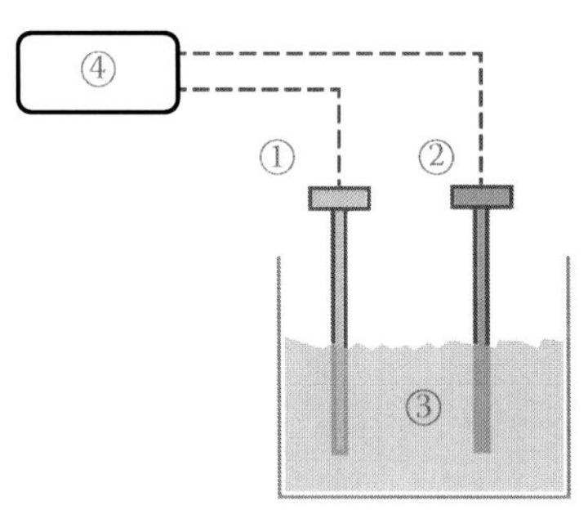

FIGURE 4.4
Temperature calibration principle.

have the master calibrated to a traceable standard with a small uncertainty. To be sure both sensors are exposed to the same temperature during the calibration, it is required to have them in a liquid bath or in a solid block with an even temperature distribution. Such baths and blocks are available from various suppliers of calibration equipment. A simple, but still accurate, bath and reference in combination can easily be made by anyone. To do this, you put crushed ice and cold water in a thermos. After some time, when the temperature is stable, you have a bath with 0 °C within less than ±0,1 °C. When all ice has melted, you need to make a new mix (Figure 4.4).

Sensors need to have a good thermal contact with the liquid in the bath (easy) or the material in the block calibrator (tricky). The block is often supplied with connection pieces in different sizes, and it is then important to select one with correct diameter, matching the temperature sensors. Also it is important that the sensor is completely immersed. If some part of the sensor stays above the liquid surface in the bath (or above the top of the block), then it is hard to be sure that the sensor is reading the correct temperature. If installed correctly, both methods are very precise and the remaining uncertainty due to temperature variations is minimised.

5

Level

Basics

Most of the tanks used in the process industry are equipped with level meters. However, in most level meter applications, it is not the 'level' that is wanted, instead the 'volume' of a stored product. An exception is when the level meter protects from overfilling; in this case, it is really the level information that is needed. When selecting a suitable and good measuring principle for level, among the first things to check is how the tank is designed.

Contactless meters (such as radar and ultrasonic) are popular. These types of instruments are mounted above the liquid surface, normally in the tank roof. In most applications, it is the liquid level that is wanted, but in this case the meter will measure the distance from itself to the surface. To recalculate this value to level, the tank height is needed. This is a value that is set in the configuration of the meter and, of course, the accuracy of the level measurement will not be better than the given height. It must be observed that many tanks will change their shape when filled, normally by an increase in diameter resulting in a reduced height (the meter will be closer to the bottom when the tank is full). One way to improve this situation is to mount the meter on a stand-pipe, not connected to the roof of the tank.

The measuring range (1) is limited by the upper dead zone (2) and the bottom dead zone (3) (Figure 5.1). The total tank height (4) is also required for the configuration. Some tanks have a reference point, a base level, inside the tank, and if so, the volume is often set to zero at this level. Remember that the nominal diameter (5) will change with rising level (6), and it is possible that all tank measurements will also change with temperature. The surface where the tank is standing must be stable so that the angle (7) of the tank will not change with time.

Measuring Principles

The most commonly used principles of level detection are based on pressure measurement, float position or echo detection. As always, various measuring

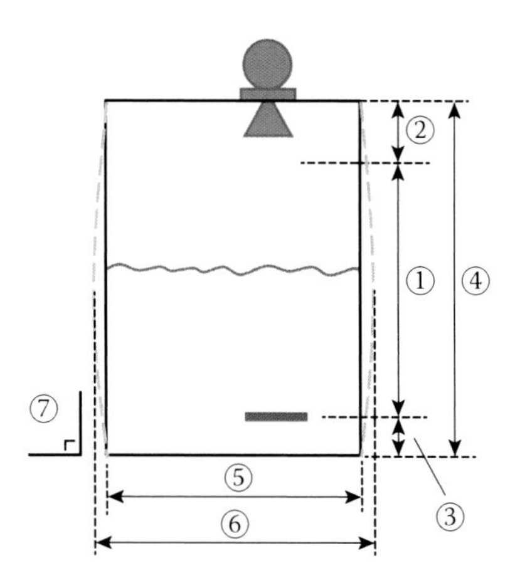

FIGURE 5.1
Tank roof-mounted level meter.

principles have their advantages and disadvantages. And, as mentioned already, if the required information is the volume, remember that the tank is also a part of the measuring instruments (a part that needs to be as stable and accurate as the meter and its requirements).

Float

A float connected to some kind of indicator is perhaps the most common type of level meter. This type of meter can be purely mechanical but will often contain an electrical switch or indicator for alarm or remote reading. The level of the float depends on its volume and weight compared to the density of the liquid in which it is floating. So, density variations will therefore cause small changes in the level indication. This 'float depth' also makes it impossible to measure all the way down to the bottom of the tank.

Ultrasonic Echo

Ultrasonic level meters are based on the time-of-flight principle. The meter will measure the time it takes for a sound pulse, generated inside the meter, to move from the meter to the surface below and back again. To be able to calculate the distance from the measured time, the meter must know the speed of sound in the tank atmosphere. Normally, these meters are configured to measure in air (internal equation will use the speed of sound in air), resulting in measuring errors if the tank atmosphere contains other gases. Temperature is also important, as the speed of sound is not constant with temperature (in air, it varies with approximately 0.2% per degree Celsius). Most meters therefore have an internal temperature sensor to compensate for variations in air temperature. For best measurement, try to make sure that this sensor is exposed to a representative temperature, as close to the average tank atmosphere temperature as possible.

Radar Meters

A radar level meter can use different measuring principles. One way is to use pulse echo (time of flight), just as an ultrasonic level meter. Another way, in most cases faster and more reliable, is to use a frequency-modulated continuous wave (FMCW), a method that uses a varying frequency. In both methods, it is the travelling time from the meter to the surface and back that is the measured value. This is then recalculated to distance, using the speed of the signal. A radar signal is an electromagnetic wave, similar to light. This speed is quite stable and varies much less than the speed of sound due to variations in gas mixture or temperature for example. To get a reading, the pulse must be reflected on the surface below the sensor. Reflexion quality depends on various factors, and one important factor is the dielectric constant of the specific media. A lower dielectricity makes the reflexion weaker which results in a smaller signal returning to the sensor. So, there is a limit where the reflexion is not good enough. This limit depends on the sensitivity of that specific level meter and the distance, but a common minimum dielectric value is around 1,5. Foam or solids floating on the surface can affect the reflexion, depending on properties and thickness of the floating material 'blocking' the sight of the surface. A table of dielectric constants can be found in Appendix.

Pressure Sensors

A pressure sensor can easily measure a liquid level, but to do so it must be connected at the bottom of the tank. There are different ways to connect the sensor: either at a hole (via a connection flange) at the bottom or by a sealed 'diving' sensor (positioned at the bottom, inside of the tank). The pressure sensor will act as a sort of weighing scale, and the measured pressure can be recalculated to level if the density of the liquid is known. With a constant volume in a tank, temperature variations will result in level differences. However, the pressure level meter will not detect this because the liquid will change its density and level correspondingly.

If the tank is closed and the upper part is not in contact with the atmosphere, there might be a different pressure on top of the liquid. This pressure must be added to the pressure measured by the level meter. For this application, a differential pressure meter should be used. If the pressure sensor is mounted at the bottom of the tank, the other side needs to be connected to the top of the tank by a small pipe (see also Chapter 3 about pressure). This pipe can be constructed to be either empty or full with liquid. As any condensate will have problems escaping, it is common to have this pipe (also called 'leg') full. This means that the output signal is inverted: when the tank is empty there is a negative differential pressure, and when the tank is full the differential pressure is close to zero. In this application, it is of course important to make sure that the pipe connecting to the top of the tank is always completely filled.

A different installation method, where the pressure at the bottom of the tank is measured with a kind of remote connection, is a 'bubbler'. A small air flow is directed through a pipe leading to the bottom of the tank. As the pressure drop in the pipe is very small, it can be neglected and the pressure measured at a T-connection at the top of the tank is considered to be equal to the pressure at the end of the tube (at the bottom of the tank). As no holes are required at the bottom of the tank, the leakage risk is minimal and the pressure sensor is not exposed to the liquid. However, this measuring method requires a constant flow of air, and in the long run, the use of air can be costly. This method will require regular service and maintenance.

Conductivity Sensor

By installing two electrically conductive plates (or bars) in a liquid, the level can be measured by means of electrical resistance. Assuming the liquid has an electrical conductivity, the resistance between the plates will decrease with the immersion depth, and thus the level in the tank. If the tank is made of a conductive material (such as stainless steel), the tank walls can replace one of the plates. The output totally depends on the installation and the properties of the measured liquid, and an on-site adjustment is always required before start up. Dirt and other layers building up on the measuring plates will affect the measurements, and this method is best where the application is clean (such as in the food and pharmaceutical industry, where this type of instruments are rather common).

By replacing the resistance-measuring device with a capacitance-measuring device, also non-conductive liquids can be measured. In other aspects, these principles are very similar.

Weighing

Weighing is a measuring technology in itself, but it is included here as an alternative to a level meter. This is because a weighing scale is normally not considered to be a process instrument, even if weighing scales, load cells and belt weighers are commonly used in many industries. Weighing scales are also commonly used in calibration laboratories and included in many traceability chains.

Basically, there are two kinds of weighing scales: a comparator type and one that measures force. In the comparator type, there is a counterweight (with a known mass) and the reading is related to this. In such a device, things like gravity (location) and room temperature are not important as it will affect both the measured object and the counterweight equally. However, most weighing scales work according to the force principle and will therefore sense not only mass. There is a difference between mass and weight. The mass is constant, but the weight will be different due to variations in gravity. Standard gravity is set to 9,82 m/s^2; however, the gravity of earth is not constant all over the planet. Due to different distances to the centre of the planet, near the equator the gravity is around 9,78 m/s^2 and near the poles around 9,83 m/s^2. However, local variations due to topography and other factors exist, and for precise weighing, it is possible to find, and compensate for, the local g value.

If using load cells under a tank, make sure that all are well balanced so that they share the total load. In many cases, three legs (or supports) under a tank is better than four. Protect the cells from side forces and temperature differences. A platform or similar, to be used for calibration weights, is good to have but it must be designed such that the weight is equally distributed on all load cells.

Air Buoyancy

Surrounding air density will affect the weight because the measured object will 'push away' (or replace) some air when it is resting on the weighing scale. In some areas, two concepts are used: 'mass in air' and 'mass in vacuum'. The true mass is the one indicated in vacuum. If weighing in air, the buoyancy, the mass of the air that the object replaces, will affect the result.

Example 5.1

One cubic metre of air has a mass of about 1,2 kg. When weighing oil (with a density of 900 kg/m^3), we should compensate for the air pushed away by the liquid. One cubic metre of oil will push away 1,2 kg of air which is equal to 1,2/900 = 0,13%. So, the air buoyance error when weighing oil is therefore approximately 0,13%.

Because the reference weight that was used to calibrate the weighing scale also pushed away some air, this should also be included (see the calibration certificate). If not taking this into account, we refer to 'conventional true mass'. As most reference weights today are made of stainless steel, we can compare the density of the object (or fluid) to be weighed with stainless steel and compensate the reading accordingly. To be as precise as possible, of course, we shall also compensate for the reference weights.

Example 5.2

If we want to calibrate a weighing scale that can be used to measure one cubic metre of oil, we need a reference weight of about the same weight. One cubic metre of stainless steel has a mass of around 8000 kg. One cubic metre of air has a mass of about 1,2 kg. When weighing a steel block, we have an error of 1,2/8000 = 0,02% because of the air that the steel is pushing away. When weighing 900 kg of oil, we would probably use a smaller weighing scale (and smaller reference weights), but the difference in percent is the same no matter the size. If using the same weighing scale to measure oil (with a density of 900 kg/m^3), we should compensate for the air pushed away by the oil. One cubic metre of oil will also push away 1,2 kg of air, but as there is a difference in density, the error in percent will now be different: 1,2/900 = 0,13%. However, probably the initial error using stainless steel weights was eliminated by adjustments performed during the calibration. If we remove this, the result will be 0,13%–0,02% = 0,11%. So, the error when weighing oil is therefore approximately 0,11%. This is also the difference between mass in air and mass in vacuum. If being precise, we should also consider that air density will change, and we can make the adjustment based merely on densities.

$$\text{Real mass} = \text{indicated mass} \times \frac{\left(1 - \frac{\delta_{air}}{\delta_{ref}}\right)}{\left(1 - \frac{\delta_{air}}{\delta_{obj}}\right)} \tag{5.1}$$

where δ_{air} is the air density, δ_{ref} is the reference, calibration weight density, and δ_{obj} is the density of object, or liquid.

References and more information regarding weighing can be found at www.oiml.org.

Level Meter Overview

A selection of level meters include pressure (1), radar (2), guided radar (3), ultrasonic (4), pressure (5) and magnetic float (6) (Figure 5.2).

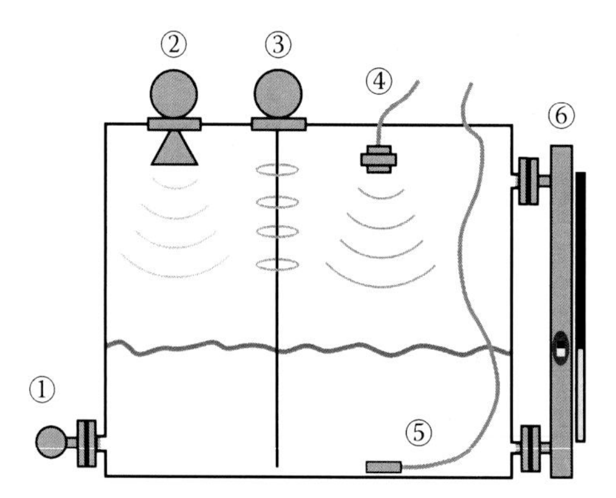

FIGURE 5.2
Various level meter principles.

Calibration and Verification

A level meter can easily be calibrated in a test rig, during a 'dry' calibration using length (a measuring tape or a laser) as reference. However, as mentioned in most cases, it is the volume that is required. So, there are two things to calibrate: the level meter and the tank. There are two common ways to calibrate a tank: geometrically by measuring the height and the diameter and by comparison with a known volume (i.e. by filling the tank via a calibrated flow meter). If the tank calibration has to be valid over time (and at various liquid levels), the tank must be robustly constructed. To secure this, a verified (or type approved) tank design can be selected. For more information regarding this, see the International Organisation of Legal Metrology (OIML) recommendation 71 or American Petroleum Institute (API) standard 620. Many level meters can be configured with the tank calibration data. Values are entered in a table, to make it possible for the meter to recalculate and linearise the measured level to volume.

6

In-Line Analysis

Chemical Measurements

There is not any clear distinction between physical and chemical measurements, but it is a good idea to separate these as there are some differences. One difference is that 'broad sensitivity' (a sensor reacting on many parameters) is considered to be good for a chemical sensor but bad for a physical sensor (in which sensitivity to other parameters than the desired is considered as disturbances). Another basic and quite important difference is how references are made. Measurements related to physics are (or at least used to be) based on an artefact, like length, time or weight, while those related to chemistry are often based on comparisons of things like colour or electrical response. This means that reference materials are needed, and this is the most common way to create traceability in a chemical laboratory. As reference materials are used with a similar purpose when performing a calibration, an uncertainty statement for each material is needed. Most materials are delivered with this information even if the terminology used can sometimes be different. There are similar organisations as for the physical area available, monitoring the work and offering accreditation for chemical measurement laboratories. For more information, visit Eurachem (www.eurachem.org) or any similar organisation.

One example of how traceability can be arranged is conductivity. As conductivity is a value of electrical resistance in a material (liquid), a typical 'physical' approach would be to relate a conductivity meter to length (area) standards and electrical standards. However, normally chemical laboratories do not achieve their traceability like that; instead, they use reference liquids with a stated conductivity. Different approaches are combined with different methods and a slightly different terminology. For the final result, this is not important, but it can create confusion when the 'physics engineer' is in a discussion with the 'chemical engineer'. Good to know when an in-line sensor is not in agreement with a laboratory test!

Many chemical measurements are made in a laboratory. For all laboratory tests, there is a need to be careful when taking the sample. If the sample is not representative of the total production, it does not matter how accurate the measurement is. Both location and time may be of importance, and an average of several samples is always better than one!

pH

Ion activity (pH) is an important factor in many processes, indicating, for example, the rate of a chemical reaction. In the food industry, pH can indicate taste, and also it can be a quality marker for a liquid and show whether the rate of corrosion is safe. The most common way to measure pH in the process industry is to use an electrode containing an ion membrane made of glass. This construction will generate a voltage proportional to ion activity. It is important that the electrode is clean so that the membrane is functioning. Regular cleaning and 'regeneration' is needed – a special requirement for pH electrodes. Frequent recalibration is recommended.

The pH scale is logarithmic and covers a very large area. So, the real activity difference at, for example, pH 7 and pH 8 is quite large. The scale spans from 1 to 14, where 7 is neutral. Solutions with a pH less than 7 are acidic, and solutions with a pH greater than 7 are basic. Examples of substance and their pH can be found in Appendix.

Conductivity

Electrical resistance in a liquid is, just like pH, a useful parameter when checking quality or when trying to separate one product from another. It can also be a safety parameter, for example, when the liquid is in contact with electrical components. The SI unit for conductivity is Siemens per meter, but other units are also quite commonly used. One of these is mho, ohm written backwards. In specifications, often S/m is replaced by µS/cm, a more practical unit for commonly used substances. Tap water has a conductivity of around 200 µS/cm, but the value is heavily dependent on the local water composition.

Special attention must be paid when dealing with liquids having very low conductivity. Just as when measuring on components with high resistance, the measuring circuit gets very sensitive to disturbances. And liquids like ultra-pure water, having very low conductivity, will also react with ambient air causing conductivity to change with time.

Temperature will be important as conductivity will change if the liquid is heated or cooled down. Therefore, readings of conductivity may be corrected to a 'standard' temperature. This procedure (manual calculation or using a built-in function) requires knowledge of the liquid and its properties. If the liquid to be measured is an unknown mix, a mathematical temperature correction will not be possible! To cool down or heat up the sample to a standard temperature is of course much more difficult, but in this case also safer.

Electrical Conductivity of Water

You will probably have heard that water and electricity make a dangerous pair together. However, pure water is actually an excellent insulator that does not conduct electricity. Pure water mainly exists in laboratories; in everyday life, we do not come across any pure water. Water can dissolve more things than any other liquid, and almost no matter where the water comes from, it will contain lots of dissolved substances, minerals and chemicals. Salt, such as common table salt (NaCl), is the one we know best. In chemical terms, salts are ionic compounds composed of cations (positively charged ions) and anions (negatively charged ions). In solution, these ions essentially cancel each other out so that the solution is electrically neutral (without a net charge). Once water contains these ions, it will conduct electricity.

Turbidity

Turbidity tells you how clear a liquid is. It is an optical characteristic and is an expression of the amount of light that is scattered by a material in a product when a light is shined through it; the higher the intensity of scattered light, the higher the turbidity. Materials that cause products to be turbid include clay, silt, finely divided inorganic and organic matter, soluble coloured organic compounds and microscopic organisms. Turbidity is a measure of how cloudy or opaque a liquid is, often a quality indication for food and beverages.

Viscosity

As already mentioned in Chapter 2, the viscosity of a fluid will influence flow profile, pressure drop and other reactions. Viscosity is the inner friction of a fluid, often in everyday life called 'thickness'. Viscosity will change with temperature, and most liquids will flow quite a lot easier (will get a lower viscosity) at higher temperatures. Newtonian liquids are said to have a viscosity independent of flow rate, and a non-Newtonian liquid will change its viscosity at higher flow rates. For these types of liquids, viscosity can go both up or down at higher flow rates. Most viscosity meters cannot measure viscosity at various flow rates and can therefore not be used for non-Newtonian liquids. Gases and most pure liquids are in general considered to be Newtonian.

There are two types of viscosities: (1) kinematic and (2) dynamic. The kinematic viscosity is the dynamic viscosity divided by the density (Equation 2.3).

The SI unit for kinematic viscosity is m^2/s, and another common unit is centistoke (cSt) (1 cSt = 1 mm^2/s). The kinematic viscosity of water at 20 °C is around 1 cSt. The SI unit for dynamic viscosity is Pa s, and another commonly used unit is centipoise (cP) (1 cP = 1 mPa s). The dynamic viscosity of water at 20 °C is around 1 cP.

Density and Concentration

The density of a gas or liquid tells us the mass of a volume unit; for example, water has an approximate density of 1 kg per litre. Concentration is a way to express how much of a specific substance there is in a mixture. It can be a mixture of two (or more) different liquids or content of solid particles in a liquid (or vice versa). If two substances with different density are mixed, the concentration can be calculated from the measured density. If there are three or more substances, or if both substances have the same density, other analytical methods must be used to measure the concentration. There are various ways to measure density; vibration is one (see also about the Coriolis flow meter in Chapter 2). A pycnometer is a density-measuring device consisting of a small bottle with a known volume. The device often has a cork (or cap) with a small hole where the liquid will be pressed out, and this will guarantee that the pycnometer is completely full. If it is then weighed empty and full, the density of the liquid can be calculated. A hydrometer is another device for density measurement. This is a kind of float with a scale on top. The hydrometer will float in the liquid to be measured, and the density will cause the float stay at various depths. The density is indicated where the scale crosses the liquid surface. Such an instrument is often used for wine and beer to measure the alcohol content. Alcohol content is often presented as a percentage, and it should be observed that this number can be related to both volume and mass. Since the density of the drink is not equal to 1, the percentage will be different. As an example, a specific beer has an alcohol content of 4,5% by volume. This is approximately equivalent to 3,6% by weight. Stronger drinks like vodka can in some regions have an alcohol content labelled 'US proof'. This old unit is still used, and an alcohol concentration of 100 US proof equals 50% by volume. Density examples of some common fluids and materials are listed in Appendix.

The term 'specific weight' is a concept similar to density: weight per unit volume. In general, weight and mass are equal, but weight is rather a force and as such related to measuring issues. 'Specific gravity' is another concept related to density, now comparing to a reference (most often water for liquids and air for gases). So, a liquid with the specific gravity of 1 has the same density as water.

Percent is equal to parts per one hundred. Small concentrations are sometimes expressed as parts per million (ppm) or even parts per billion (ppb). Here, it should be observed not only that volumes and weights have different units across the world, but also that the concept of a billion is different in areas across the world and should therefore be avoided. To be clear, we must also add if the stated part is by mass or by volume. Another common way, mainly used for solid particles in gas, is to express concentration as mass per unit volume. Then, it is also good to state a standard condition, like 3 mg/m^3 @ 102 kPa and 20 °C. The @ symbol is then used with the meaning 'at'

100% = 1 000 000 ppm = 1 000 000 000 ppb (in the USA and Europe)

7

Electrical Signals

All measuring instruments have an indicator to show measured data. It can be a local display for your eyes to observe, or an electrical signal for remote reading. Instruments used in process industry almost always have one or more output signals. Analogue electrical current signals (mA) are frequently used. There are a couple of standardised ranges where 4–20 mA is the most common; 4 mA equals 0% and 20 mA equals 100% of the process value (e.g. 0–100 °C). One can say that the 4–20 mA signal is a universal percent indicator, and you must always (in most cases manually) set ranges and units at both ends (sender/measuring instrument and receiver/indicator). It is important that both devices are set to the same values, failing to do so can result in large measuring errors.

Some instruments will instead, or in addition, send pulses as an output signal. This method is common when measuring an amount such as tons or cubic metres. The pulses are in that case sent to a totaliser that will count the number of pulses and indicate the total amount. In this case, you need to set pulse value in both sender and receiver; if the meter will send one pulse per cubic metre of course, the totaliser needs to be set to indicate cubic metres.

A newer method is to use data communication. There are a few data bus systems specially developed for process industry environment and needs, such as PROFIBUS and Fieldbus Foundation (http://www.fieldbus.org/). M-Bus (Meter Bus – designed in accordance with EN13757) is similar but focuses on utility metering (water, gas and electricity) rather than process metering. Network devices such as Ethernet and Modbus, normally found in office systems, can also be used as long as environmental conditions allow. One major advantage of using data communication is that more data can be sent through the same wires. Several measurement values, including range and units, are easily transferred which reduce the risk of errors in settings. Also, error messages and alarms can be distributed and some systems will also allow for remote service and configuration. With digital data communication, signal quality is easier to maintain – if the communication is up and also running, transferred data are correct. However, all basic settings for the network must be correct and drivers must be updated, which is not an easy task to perform. If connected to Internet, it is very important to observe the risk of hackers and cyber attacks on systems and networks. ISO 62443 provides more information on how to protect control systems from cyber attacks.

A quite common system is HART, where an analogue signal is combined with a (slow) data communication. HART can transfer measurement data but is mainly used for configuration purposes. Remote configuration by a HART hand-held terminal (or a PC with a HART modem) is a convenient way to access instruments installed remotely. To use this type of configuration, tools also offer the possibility to archive settings in an easy way. Also in some large control systems, HART functionality is supported. The term 'distributed I/O' refers to a control system where input and output ports are located in external units, located close to the process devices. Between these units and the central control system, the communication is digital, and from the sensor or actuator, analogue signals are used. Also here, HART functionality may be possible.

Signal Transfer

Inside almost all measuring instruments, you will find several signal converters and calculators. Also in the receiver (controller, display or system), there are converters. Finally, you can sometimes find external converters and protection barriers between the two. All conversions must be done correctly and it is the job of the instrumentation technician to make sure everything in this chain is correct. Extra care is needed for all those converters with a settable (or programmable) configuration, as this can result in large errors if not set correctly.

Figure 7.1 describes how music is recorded, stored and distributed. However, the same structures are used for many measurements, and in the

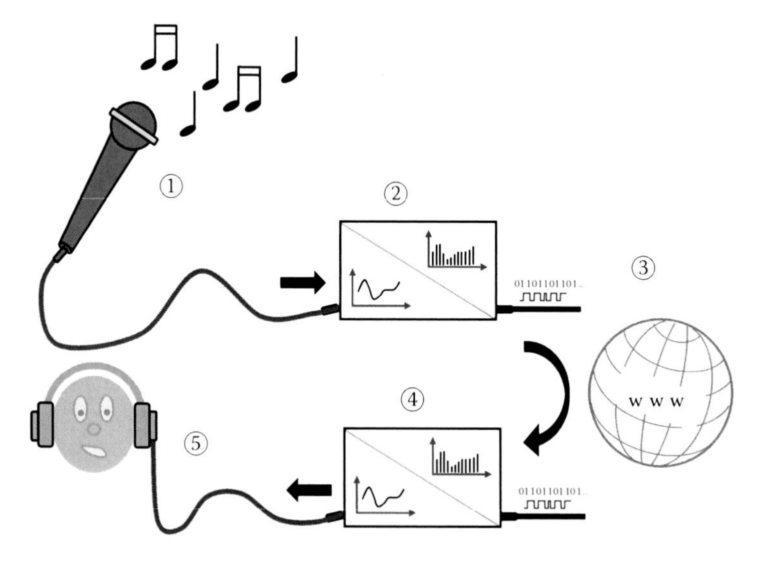

FIGURE 7.1
Analog to digital to analog signal conversion.

process industry similar systems are built. First, there is a sensor (1) collecting data. The data are fed to a signal converter (2), where in most cases, analogue signals are also converted to digital information. In some cases, this converter has a built-in function also to generate an analogue output; in other cases, it will transmit digital information to a network (3). This network can be local or public, because of safety and security reasons most process industries are not connected to the Internet. The receiver (4) is then another client on this network, collecting selected information. In the case of a process industry, it can be a specific valve (5) searching for tank-level information or similar, capable of taking actions (open, close or stay) to control this process parameter.

Analogue Signals

$$I = \frac{V}{R} \tag{7.1}$$

where I is the current (A), V the voltage (V) and R the resistance (ohm).

For measuring an analogue signal correctly, you need some fundamental electrical knowledge. Ohm's law will assist you in understanding both possibilities and limitations of the signal. When looking at this, you will quickly understand the advantage of using current instead of voltage for signal transmission. A current will remain at the same level all the way through a loop, no matter the cable length (or if there are terminals and connectors along the way). If the total load (resistance) in the loop is not above maximum, the signal value will be correct wherever along the loop it is measured. All cables have some resistance and at some point, the cable will be too long. But if you stay below the maximum loop resistance, the mA signal will be OK. If you go above the maximum resistance, the signal will be too low. However, a voltage signal would drop constantly, and the distance from the sender to the receiver would cause increased resistance and error in reading. Therefore, when using voltage signals, all cables should be kept as short as possible!

A special version of the current signal is the 'two-wire' loop. This type of wiring is also called 'loop supply' because the supply energy to the measuring instrument is sent along the same wires as the measuring signal. This can be achieved when there is enough power to supply the instrument also at 4 mA, so that the instrument can also 'stay alive' when the measured value is zero. This method is popular because it reduces the need for electrical wires and it makes installation work easier. However, the power to operate the instrument is limited to a rather small value and not all instruments can be energised like this. Special attention must be taken to the maximum load in the loop, as here also other components will consume parts of the available power. There are also instruments with a mix of systems, where output signal is connected as a two-wire loop and where there is still a need

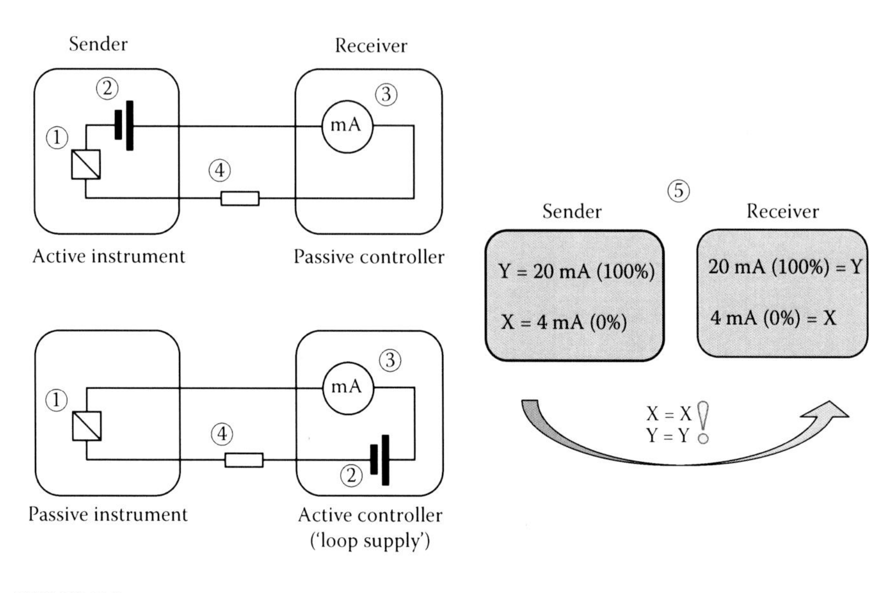

FIGURE 7.2
Active and passive analog signals.

for a separate power connection. This results in a common question for all 4–20 mA terminals: Is the signal output active or passive? You need to know this before connecting a receiver or an indicator: an active output will require a passive indicator and a passive output will require an active indicator. Figure 7.2 shows that the main difference is the location of the power supply (2). The measured value generator (1) as well as the indicator or receiver (3) remains at the same place. The total resistance (4) in the loop must in any case not exceed the maximum limit. When the connection is ready, you must also make sure that the configuration is correct. The measuring range is in most cases settable in both sender and receiver, and both ends must be set to the same values.

Most control systems somewhere have converters that make digital data of analogue signals (A/D-converters). An important specification for each A/D-converter is its resolution. This denotes the size of steps between each possible analogue value. The binary term 'bits' is often used to specify the resolution, where 9 bits equals an approximate step size of 0,2%, and 12 bits is better and here the step size is about 0,02%. In most applications, 16 bits is very good with a step size of 0,002%. A low-resolution A/D-converter somewhere along the signal chain will destroy all good accuracy data. High resolution is almost always better, but it can result in a slower operation (Table 7.1).

Common problems in mA-signal circuits are faults in the ground (or earth) connection, especially if more than one receiving instruments are connected in series. Some devices have no galvanic separation between signal zero (reference) and supply zero (reference), and this can result in problems.

TABLE 7.1

Binary Numbers versus Analogue Resolution

Bits	Resolution (%)
1	50
8	0,4
12	0,02
16	0,002

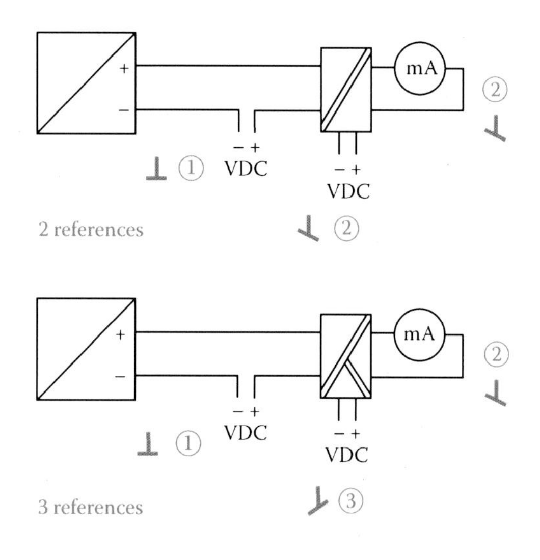

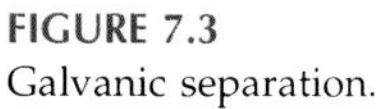

FIGURE 7.3
Galvanic separation.

A simplified explanation is that the current 'takes the wrong path' and if this happens, the bypassed indicator will indicate zero or a very low value. In many cases, this problem can be solved by just connecting the devices in a different order, so that the 'grounded' indicator is the last device in the loop. Another solution is to use an 'isolator' or 'i/i-converter'. An isolator will copy an incoming signal to the output but with a new (isolated) reference, which will prevent the system from grounding/earthing problems. Isolators with two references will separate input (1) and output (2) from each other. An isolator with three references will also keep the supply voltage (3) separated (Figure 7.3).

HART

HART (Highway Addressable Remote Transducer) is a commonly used protocol in many process measuring systems. HART means data transposed on top of an analogue 4–20 mA signal. By this method, you can transfer a

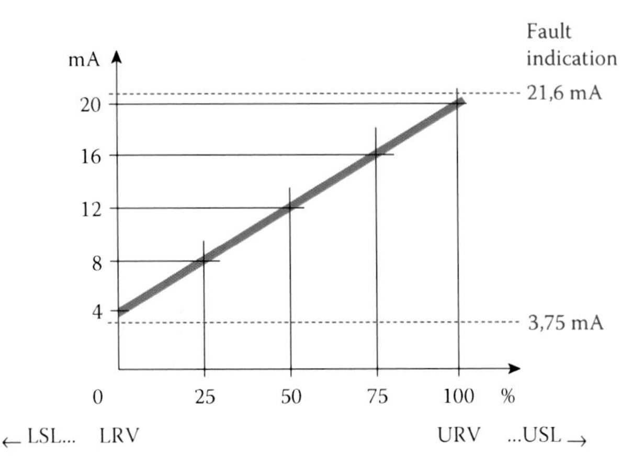

FIGURE 7.4
Signal range.

traditional measuring value (such as 0–100 kPa) and, in addition, transfer settings and check diagnostic values in the instrument. The connection can be done with a special hand terminal, a device with integrated HART modem and software, or by using a PC through a HART modem. The software can be either a 'one-to-one' communicator or a more complex system where the software is capable of connecting to a network of many sensors. PACTware™ is an example of network software, supported by many different manufacturers.

Communication using HART can be done in two levels. There is the generic mode, which does not require any special drivers. In this mode, you will not be able to connect to all functions in the instrument, only basic settings like range and zero adjustment are useable. In the full mode, you need the specific driver (Device Description, or DD file) for that instrument you want to connect to. If you have this file installed in your terminal or PC, you will be able to connect to all functions and settings. Important settings in the configuration include the measuring range, limited by the lower range value (LRV) and upper range value (URV). Minimum LRV is device-specific and limited by the lower scale limit. Maximum URV is limited by the upper scale limit (Figure 7.4).

For more information, please visit en.hartcomm.org and www.pactware.com.

Pulse/Frequency Signals

As mentioned, pulses are in most cases used to transfer measured quantities (integrated data) to counters or totalising displays. However, the frequency of pulses can also indicate a measuring value, for example, where 0 Hz is equal to 0% and 1000 Hz equals 100%. In many instruments, it is therefore possible to configure the pulse output in two ways, either by setting a pulse value (such as 1 kWh per pulse) or a frequency range (such as 0–1000 Hz is

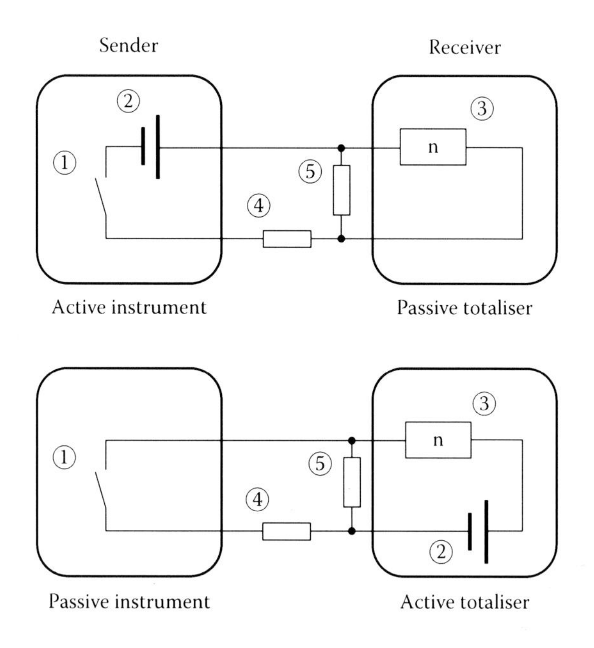

FIGURE 7.5
Active and passive pulse signals.

equal to 0–200 kW). However, this is not always possible; the design may be such that the frequency is unstable during operation, and if so, this will cause an unstable reading (the total number of pulses might still be correct).

When designing a pulse system the frequency and the number of pulses are important but impedance and voltage levels must also be checked. The instrument itself might have two different types of output. It is common to use a switch, which closes for every pulse. In this case, the voltage supply must be arranged externally. In other instruments, an active voltage signal is generated at the pulse output, supplied from the inside of the instruments. Of course, the receiving instrument must fit to the output available. Resistance is also important; if resistance is too low it wil cause overload (a too high current) and if it is too high it will cause high sensitivity to signal disturbances.

Figure 7.5 shows both active and passive systems and as you can see they are very similar, just as for analogue systems the main difference is the location of the power supply (2). The pulse generator (1) can be a mechanical switch, but an electronic relay (or transistor) is more common. Also, the totaliser (3) are usually an electronic device with LCD or LED indication. The total resistance (4) and the voltage of the supply are limiting factors for the current in the loop. If the internal resistance (or impedance) of the totaliser is high, an external resistor (5) can prevent from counting errors due to electrical disturbances while the switch (1) is open.

A digital signal, for example a limit switch, is very much like the pulse signal described here and the electrical connection is identical. There are some frequently used concepts that can be good to know. In Figure 7.5, the contact has two accessible connectors, and it can be either open or closed. Because it is drawn 'open', its normal position is in 'off' state. There are other types of contacts switching between two positions, these having three terminals where one (common or 'C') switch an electrical current from one terminal (normally closed or 'NC') to the other terminal (normally open or 'NO') when activated.

In the abbreviation SPDT, SP refers to a contact with a single pole (like the one described above). SP and DP refer to single pole and double pole, respectively, and ST and DT refer to single throw and double throw, respectively. Pole refers to the number of circuits controlled by the switch meaning that an SP switch can control only one electrical circuit. A DP switch controls two electrically independent circuits but they will switch at the same time. Throw refers to positions of the contact. ST switches close a circuit at only one position. The other position of the contact is off. DT switches close a circuit in both 'active' and 'non-active' position (on–on). An SPST contact normally has two terminals, whereas a DPDT contact is supplied with six terminals.

Controllers

One of the main purposes for industrial measuring instruments is to control a process. In such cases, the measured value, the output signal, is connected to some kind of actuator. It can be a valve controlling the flow, a heater to control temperature, and a pump or something else that can 'influence' the process. Flow, temperature, pressure and other process values are then used to create (or adjust) various process reactions, such as mixing, quality and similar (as described in Chapter 1).

In basic terms, the controller is quite easy to understand. You can compare with yourself adjusting the temperature in your room. If it is too hot, maybe you will open the window (now we assume you have no air-conditioning available). Probably, you will first open it wide to get as much fresh air as possible. But then, it quickly gets too cold. Thus, now you close it a bit so that the room will get comfortable. The controller works in a similar way. To optimise a controller can be a rather complex procedure. You need to know signal ranges and process time constants. When the controller is active, there are mainly two values that it needs to know: What is the actual process value (measured variable, e.g. the pressure, or the temperate in your room)? What is the desired process value, the set point (the required pressure, or the desired temperature of your room)? If the difference between these two values is zero (or within a small difference), the controller will keep the output at a constant level.

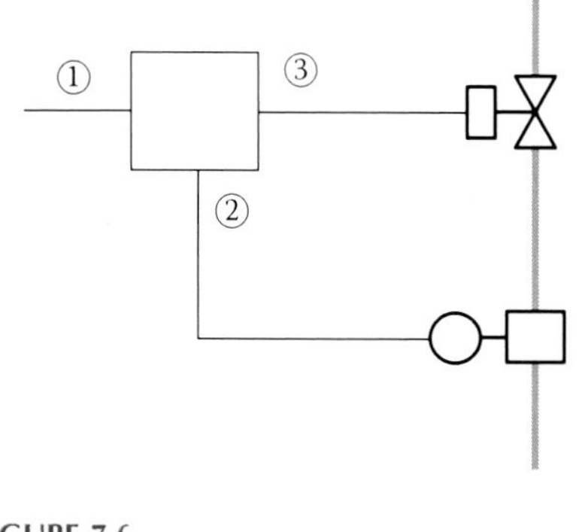

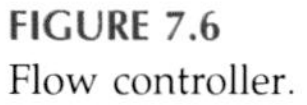

FIGURE 7.6
Flow controller.

If a control technician and a measurement technician talk about 'errors', they often refer to slightly different definitions of this term. Error in a measurement system is the difference between the measured value and the true value. In a control loop, error is the difference between the set point and the measured process variable. Of course, misunderstanding can cause problems! It is good to know that there actually can be three values around a controller: set point value, measured variable value and true variable value. This is often overlooked in control system courses. There is also another aspect when discussing control errors. It can be that the actuator is too coarse, for example, a control valve that is too large. This problem will result in a situation where you know that you have, for example, a mixture that is not correct. The controller can detect the deviation but cannot adjust it because of lack of precision in the actuator, valve or similar. In a control loop where low uncertainty is required, it is not only the measuring instrument that should be observed but also valves, actuators, heaters, motors and all components in the loop and, of course, the controller itself!

Explaining a little bit in detail, it is not possible to connect the measuring instrument (2), the process value, directly to the actuator (3), we need the controller in between (Figure 7.6). A device adjusting signal levels and adapting these to the characteristic of that specific actuator is required, as is the requested value (1): the set point. In other words, to compare the process value with the set point and then generate an output signal to the actuator according to the deviation (or control error) is the job of a controller. There are several different types of controllers, that work using different principles, but two common types are the 'on/off controller' and the 'PID controller'. The first will, for example, switch on heat at full power when temperature is low and switch off when the temperature is high. This will result in not only a controlled temperature but also a constantly varying temperature. The PID controller can adjust power between 0% and 100%, and it also has the capability to compensate for time delays in the process and the actuation time. With a PID controller, correctly tuned to the specific process, the temperature is controlled to a stable level.

For more information on control technology, there is a lot of literature available, for example, *Introduction to Process Control* by Jose A. Romagnoli and Ahmet Palazoglu (CRC Press). You can also browse the book store at www.isa.org.

Calculations

In many cases, the measured signal needs to be transformed, copied, multiplied or added to another signal. In all these cases, a signal calculator is needed. Just as the controller function, this calculation can be performed in a small module or in a 'block' (or function) in a control system. Next you will find some common examples of calculating functions (Figure 7.7).

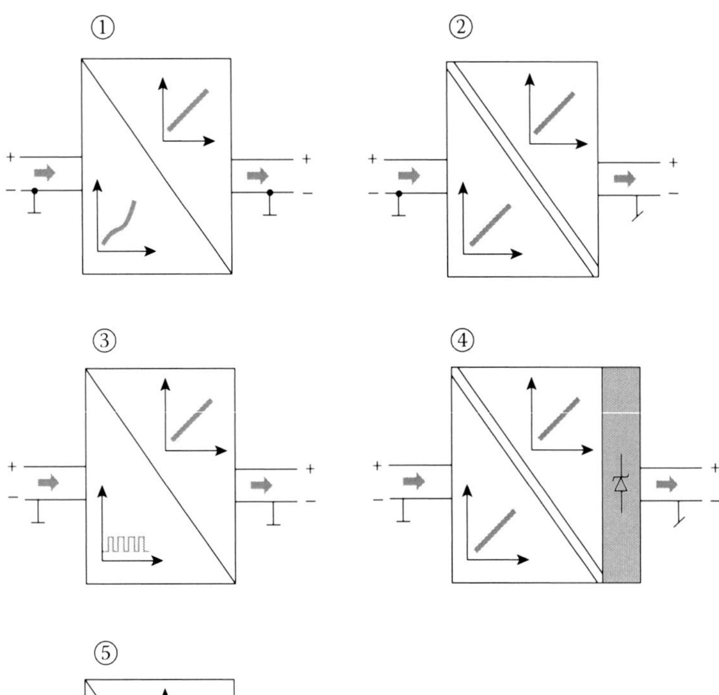

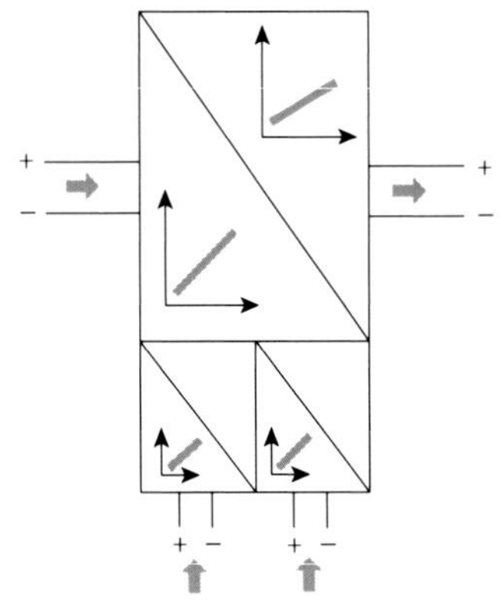

FIGURE 7.7
Various signal converters.

Linearisation

Linearisation (1) is required when a sensor produces a non-perfect signal. If the signal is always showing the same relationship between the input and the output (if the sensor is repeatable), the final result will improve a lot with a linearisation in between the measurement and the user. In practice, there are some different ways to set up this. The most common is to use a table with various numbers of points and linear correction between these points. Of course, you need to know the 'true' relations before you can set up a linearisation table. See the chapter on calibration for more information regarding this. Pure mathematical linearisation, such as square root, is also common, especially when working with differential pressures.

Galvanic Separation

Galvanic separation (2) is a sort of recalculation as well, even if the conversion is one to one. The signal coming in is also going out, always with the same value. The purpose of this commonly available device is to isolate one signal ground from another. This will improve the protection from disturbances, and it can simplify the connection of several units in a large network of sensors. When it comes to calibration and control, you must make sure that the isolator is accurate enough so that a small change in signal level from the sensor will not be lost in the isolator. Always include the isolator in the calibration!

Pulse to Analogue Conversion

Pulses normally indicate total amounts and, therefore, it is the number of pulses that is equal to process data. However, *frequency* can also be measured on the same signal. In an energy meter, number of pulses can be converted to energy (kWh) and frequency to power (kW). The frequency can be externally converted to an analogue signal, useful in a controller for example. This is done in a D/A-converter, or an f/i-converter (3), which is a more specific device to transform Hz to mA. The conversion can be done using two different methods, either by counting pulses within a given time or by measuring time between pulses. The first principle will give the best results at high frequencies, and the second method will be best for low frequencies. There are devices that can automatically switch between both methods and, of course this is the best, if the signal will vary a lot in frequency.

Ex-Barriers

The purpose of the ex-barrier (4) is to prevent dangerous amounts of electrical power in signals from transmitting into a potentially explosive atmosphere. By limiting the power, there is not enough energy to get an ignition by the spark that can be created if there should by accident be a short circuit somewhere along the cable. To indicate that the signal (and cable) is protected, it is common to use blue cables and terminals. However, even if output energy from a barrier is limited to a safe level, it is not possible to connect any device. If someone, for example, would connect a battery to the signal cable from a barrier, it would negate the safety because after some time the battery has collected enough energy to be able to create an ignition. Therefore, every ex-barrier is labelled with maximum capacitance and inductance in connected equipment to avoid the 'battery effect'. Basically, ex-barriers are a kind of galvanic separators and one-to-one converters. Several functions may be combined into one unit resulting in, for example, a linearisation device with a built-in ex-barrier.

Heat and Steam – Thermal Power

A heat meter (5) is made to measure thermal energy. Here, the word 'thermal' is used to tell the difference from an electrical energy meter (the most common type and normally what is referred to if someone talks about an energy meter). These instruments are designed to measure energy, but many of them also have functions to indicate power. A heat meter needs at least three sensors, as thermal power is calculated from flow and two temperatures. The output is calculated from all three inputs. Normally, it is the calculator that is called heat meter, but as flowmeter and temperature sensors are always needed, it is the combination that is the heat meter. Energy is calculated based on volume (or flow rate and time), temperature difference between inlet and outlet, and enthalpy (energy content) of the media in use (usually water). As a rule of thumb, 1 L of water contains 4,2 kJ per degree Celsius.

Thermal heat can also be distributed as steam. The energy content in steam will vary with pressure and temperature. Let's consider an example:

Even cold liquid contains energy. Water as an example contains around 4,2 kJ per kg and degree Celsius (or Kelvin). This means that if you add 4,2 kJ of energy to 1 kg of water, the temperature will rise 1 °C. If you then add another 4,2 kJ, the temperature will rise one more degree. This will continue up till you reach 100 °C where the water will start boiling. Now, some water will transform into steam and leave the liquid. If you imagine that both water and steam are enclosed in an expandable tank, we still however have 1 kg of water: some in liquid phase and some in gas phase. If we continue to add energy, more steam will form, but temperature will remain at 100 °C until all liquid is gone and we have 1 kg of steam. This steam is called saturated steam. At the same time, the volume has now increased from 1 litre to around 1600 litres. If we started with water at 0 °C, we have now in total added around 2,5 MJ heat. If we continue to add even more energy, we will make our steam superheated.

If we know that the steam is saturated, we can use a measuring device to calculate pressure from temperature (or temperature from pressure). This is of value if we need to know density and energy, for example, in a flow meter.

However, if extra energy was added during production, and the steam is now superheated, we need to measure both pressure and temperature for calculating density and energy.

The SI unit for energy is joule (J). Energy is time multiplied by power and 1 J is equal to 1 Ws. Therefore, 1 kWh is equal to 3600 kJ. It is also true that 1 W is equal to 1 J/s.

Standard Volume

When measuring flow and volume, it is often necessary to recalculate the measured value. The purpose of this is to correct the volume for pressure and temperature, especially if it is a gas measurement as this correction will then have a large impact on the result. Using mass units instead is easier in many ways. A kilogram is always a kilogram, but a litre is not always a litre. Because of this, you will also find another related concept: standardised volume or normal volume. This is a volume recalculated to 'standard conditions'. The device that can do this is very similar to the heat meter described above, but with a different equation in use. There are not any standards telling us how to do this but still it is common practice to use these 'units', and they are often indicated with an 'N' or 'n' in front (as Nm^3). Normal pressure is in most cases atmospheric pressure: 101,325 kPa (a). Normal temperature can vary but is normally 0, 15 or 20 °C or equivalent values in other temperature scales. You must find out what normal (or reference) conditions mean in your application. Standard conditions are other words for the same thing.

When trading commercially with oil and gas, correction to normal conditions is a standard procedure. Sometimes, the recalculation will not stop at normal cubic metres, and the calculator will also present other volumes and sometimes also energy (based on fixed enthalpies related to a specific product). The device that can handle this is a flow computer. A flow computer contains verified algorithms, density tables and standardised correction factors. To do, a recalculation is normally not very complicated, but the fact that it needs to be done in a secure way and in a way that is possible for everyone to verify, including customers, suppliers, traders and authorities, makes it more complicated. For this reason, flow computers are often certified and form a part of the verified measuring system in a custody transfer application.

Thermal Expansion

Not only gases but also liquids and solid materials are dependent on temperature. Density and volume of the material will change with temperature. This can result, for example, in changing diameters of a pipe, or a different volume in a tank. If the liquid that is measured changes temperature between the point where it is measured and the point where it is used, recalculation and compensation might be required. In some applications, this volume (trapped between meter and user) is called *buffer volume*.

Tables and Standards

When looking for constants, physical data and similar the information you find may sometimes differ depending on the source. Examples can include steam density and energy content of gases. There are two main reasons for this, either the fluid is not specified in detail (for example in a water density table, what type of water do they refer to). Another reason can be that the standardisation bodies used different methodology to obtain the data. In this case it could not be said that it is wrong to use data from one of these sources, but it is good to mention the source. In some devices (like calculators and flow computers) it is of this reason possible to select what standard to use.

Calibration

Electrical instruments for voltage and current are useful and important tools. If used in the measurement circuit for calibration purposes, the instrument must be of good quality and calibrated. A loop calibrator is a precise mA-meter with a built-in supply capability. The loop calibrator is basically connected as a replacement of the output signal of the original instrument. When switched on, the loop calibrator can simulate different process values from 0% to 100%, and the input is compared to the reading in the system. A common and practical approach is to set up an acceptance limit; if there is a small difference between input and output, this will not result in any actions. The size of this limit must of course be small enough to, in practice, be negligible. However, when working with measurement uncertainty, this acceptance limit must also be considered.

System and Design Drawings

Process design drawings are very useful, both when designing new systems and when searching for problems. Various standards are available for this, detailing how symbols should be used for various components. There are also different types of drawings, aimed at different working groups such as mechanical design and electrical design. If working with measuring

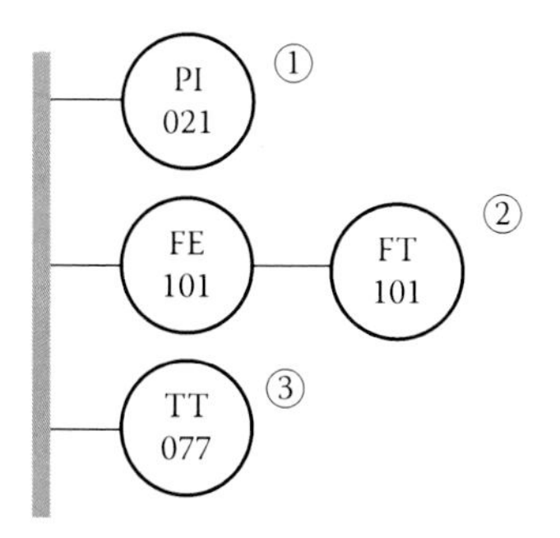

FIGURE 7.8
Identification of instruments.

systems, a very useful type of drawing is the Process and Instrumentation Diagram (P&ID). This gives a schematic overview of process flows and electrical signals (or in other words, both pipes and cables). Some of the most common symbols used in this type of drawings can be found below.

Symbols used in P&ID schematics are not standardised. There are several standards available, offering small differences. The basics are very similar, with symbols looking like the ones previously used in this book. Symbols may also be simpler, using only a circle for a measuring instrument. This can then be labelled with a name, a 'tag number', unique for that specific position. The tag number can include codes for what type of instrument it is, where P stands for pressure, F for flow, T for temperature, L for level and so on. The second letter stands for the function of the device, where I stands for indicator, T for transmitter and E for element (sensor). There are also codes for controllers, recorders, safety functions and more. One commonly used standard in this area is ISA 5.1. Visit www.isa.org for full information and complete code keys for this standard. There are also various PC-software available for making P&ID drawings and many of these have symbols built in.

In Figure 7.8, three instruments are shown. On top a pressure indicator (1), in the middle a flow meter with a separated sensor and converter/transmitter (2) and, finally, a temperature transmitter (3).

8

Valves, Pumps and Pipes

Valves

There are many different types of valves, some with a universal design and some tailor-made for a specific application. When it comes to measurement and process automation, besides usual selection requirements such as media, temperature and pressure, there are a few things to consider. For example, is the valve 100% leak free? It must be observed that not all valves even in theory completely will block the flow in closed position. When the system requires a totally closed valve, one option can be to use a 'double block and bleed valve'. Basically, this is two valves in series with a small check valve on a T-connection in between. When both valves are closed and the check valve is open, a complete closed flow (and isolation) is guaranteed. Such valves are often used in custody transfer applications.

Among often-used types of on/off valves are ball and butterfly valves. An on/off valve can also be used as a control valve, but it does not produce very good results as the setting in most cases is too coarse.

Valves should always installed downstream of a flow meter. If this is not possible, and a valve is installed upstream and in front of a flow meter, it must be ensured that this valve will disturb the flow profile as little as possible. Use a type of valve that in open position has no restriction in the pipe diameter.

Control Valves

As the purpose of most flow meters in process automation is to control production speed, level, temperature, pressure or the flow itself, control valves are very often found near flow meters. In any case, always install the control valve after the flow meter. Just as for a shut-off valve, there are several different types of control valves to choose between. The selection also often includes selectable inner parts, various sizes of cones and cylinders. This is to enable a good control function. To select the best control valve, information about the media, upstream pressure, flow range and also downstream pressure is required. To estimate the downstream pressure is difficult, as it can depend on the selected valve. The requested information is rather the required pressure for the downstream process, like the pressure drop in pipes and components after the control valve.

A sizing variable for control valves is the K-value. Basically, this value will tell the flow at a specific pressure drop. Three different K-values are used: Kv (stated as flow in litre/min at 1 bar differential pressure), Cv (stated as flow in gallons/min at 1 psi differential pressure) and Av (stated as flow in m^3/s at 1 Pa differential pressure). Even though Av is the SI unit, it is hardly ever used.

Pumps

The most common pump type is the centrifugal pump. One or several turbines rotate at high speed inside the pump and cause the liquid to move. Another common type is based on the volumetric principle, where small chambers (portions) of liquid are moved in sequence from the inlet to the outlet of the pump (positive displacement type). When it comes to measurements, some pumps (1) will create problems for a downstream flow meter (2), resulting in disturbances on the flow signal (Figure 8.1). A volumetric pump will cause pressure variations with the same frequency as the movement of the pump chambers. If the chambers are large and move slowly, like in a membrane pump, the flow and pressure variation may be possible to measure. If the frequency is high (as in a piston pump), the unstable pressure may cause unstable signals and measuring errors. In most cases, a volumetric pump will require a special flow meter, and in an electronic meter, filter functions will be useful.

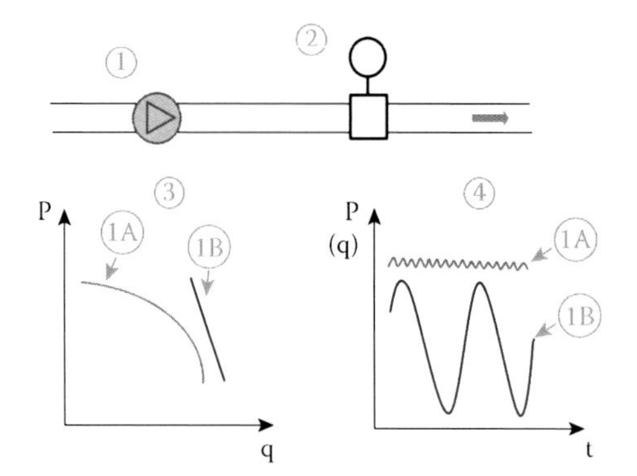

FIGURE 8.1
Flow and pressure relations.

The left diagram (3) in Figure 8 shows the simplified relation between pressure and flow in a centrifugal pump (1A) and a volumetric pump (1B). The right diagram (4) shows pressure variation in time, also for a centrifugal pump (1A) and for a volumetric pump (1B). In a static system, pressure variations will also cause flow variations. The illustration is not in scale, and actual variations depend on pump type, size and application.

Head

Head is a concept used for pumps. It is a combination of pressure and density of the fluid pumped, and it will tell you how high the pump can lift a fluid. If you add the pressure on the inlet of a pump with the pressure on the outlet, you have calculated the total head. This in combination with the flow rate and properties of the fluid is proportional to what work a pump performs, that is, power and energy. If the energy lost in the pump, drive and electrical motor is known, the efficiency of that pump can be calculated.

The negative pressure on the suction side can not be lower than 0 bar a (or −1 bar g which is equal to vacuum). In theory, that means a pump can never be positioned higher than approximately 10 m above a water surface to be able to pump. However, in practice, due to fluid, pressure drops in pipes, weather conditions and other imperfections, the maximum height will be lower. For example, when pumping water at 70 °C, the maximum height on the suction side will be around just 2 m. Pump type and fluid set the limit. In some cases, even a positive pressure might be required to make the pump work.

Pipes and Fittings

Pipe sizes are normalised. There are many standards, some general and some application specific (e.g. those used in the food industry). Each pipe and connection size is named after its nominal diameter; however, measures vary with pressure rating. For a pipe, it is the outer diameter that is equal for all pressure ratings in a specific dimension, and as the pipe wall is thicker in pipes suitable for high pressure, the inner diameter will be smaller in a high-pressure pipe. This is of importance for some instruments; for example, vortex-type flow meters need to be adapted not only to nominal diameter but also to pressure class. There are two common pipe size standards: the American (ANSI/ASME/API) standard with sizes stated in inches, and the European/German (DIN) system with dimensions in millimetres. In the American system, the pipe diameter is known as the 'nominal pipe size' (NPS) or 'nominal bore' (NB), and wall thickness is categorised by the 'Schedule'. In the European system, it is known as the 'nominal diameter' (DN). Each nominal diameter is available in different wall thicknesses, to match working pressure and temperature. Pipes can be made of different materials; mild steel, stainless steel, copper and plastic pipes are commonly available. Steel pipes can be longitudinal welded, spiral welded or seamless. In a precise measuring situation, watch out for welded pipes that are not smooth inside as these can affect flow profile. The pipe needs to be round if a clamp-on flow meter is used. Even small imperfections in roundness can affect the measuring accuracy.

Flanges

Just as for pipes, there are a few different flange standards in the process industry; ANSI, DIN and JIS (Japanese Industrial Standard, used in Asia) are the most commonly used. In all systems, there are both nominal sizes and nominal pressure ratings. According to DIN and EN, working pressure is stated by 'nominal pressure' (PN), and in ANSI, there are pressure classes in pounds (lbs). To use the same standard all over a plant is of course an advantage, as different flanges will not fit to each other. In some cases, especially for pipes with diameter less than 100 mm, flanges within the same standard but with different pressure class can still be connected to each other. However, in most cases, different number and dimensions of bolts are used and these flanges will not fit, even if the nominal diameter is the same (Figure 8.2). Nominal dimensions for some flange types can be found in Appendix.

Threaded Connections

On pipe fittings, there are three commonly used thread types: NPT, BSPT and BSPP. NPT (National Pipe Thread) is a conical connection with good sealing properties often used for manometers and pressure transmitters. BSPT (British Standard Pipe Thread) is similar to NPT but with a different angle in the conical section. In some areas, this type is also called 'R type'. BSPP (British Standard Parallel Pipe) is a parallel thread fitting that uses a ring for the sealing.

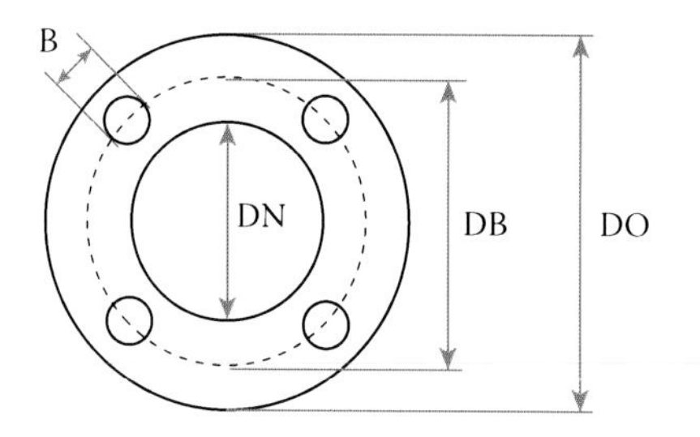

FIGURE 8.2
Flange dimensions.

Special Connections

Many pipe fittings and quick connections are available. Most of these are used in specific industry segments and designed with a special requirement in mind. Food and biochemical-type connections are often made of hygienic and polished stainless steel with seals and O-rings designed to protect from bacteria forming. Connectors used on fire trucks and tanker trucks are used for mobility and quick connection and release.

Cavitation

As described in the section on thermal power, liquids can start boiling even at low temperatures due to low pressure. Sometimes, gas bubbles are not desired. In a hot liquid, it may then be needed to keep the liquid under pressure to stop gas bubbles forming. In a pump, this can be a major problem. For example, if you have a pipe where the pressure is 2 bar g, water temperature can go up to 120 °C, and the water is still in the liquid phase. At 16 bar g, water up to 200 °C will remain liquid. As there is always a pressure drop when a fluid is flowing through a pipe, valve or any component, the pres sure will change inside this component. In some cases, pressure will also increase. During specific conditions, at first gas bubbles will form, and then just shortly after return to the liquid phase. At this return, the bubbles will implode, creating quite a lot of unwanted energy that can destroy material and cause all sorts of problems. This event is called cavitation, and if it occurs, you could easily hear the noise caused by these implosions. In a pressure-reducing device, like a control valve, it is better to do the reduction in several steps as this can prevent cavitation.

9

Safety

There are many different aspects of safety in a process industry. As this book is about measurement, we will here focus on safety related to measuring equipment. Most safety systems include a measurement function and, therefore, safety issues may be an important part of the measuring instrument, or the 'sensor' as it is called in some safety-related standards. Basically, we can divide all safety issues into two parts: The first is everything related to preventing the instrument itself from causing accidents or injury, for example, protection from electrical shock or leakage. The second part relates to functional safety, where system safety or other components rely on correctly measured data.

As safety is always important, it is common that safety aspects of an instrument are assessed by an independent organisation. If so, there are reports and certificates that can be used to verify the safety on site. The need for safety features (and corresponding documentation) depends on the application and risks associated with the specific process. Normally, national laws issued by governmental bodies set the requirements. However, insurance companies and other organisations can issue recommendations, and often the company itself will set up internal requirements, or tolerances to certain risks. This can be referred to as a company's 'risk appetite': How much risk of injury or death to an employee or member of the public can the company tolerate?

This chapter provides a bit more about some common safety issues, related to measuring instruments.

Electrical Safety

Worldwide governmental organisations have published recommendations for electrical safety. In the United States, such documents can be found at www.osha.gov [published by Occupational Safety and Health Administration (OSHA)]. European recommendations are similar and can be found at all member states, for example, at www.hse.gov.uk/electricity/ [published by Health and Safety Executive (HSE)] in the United Kingdom. Inside the European Union, the Low Voltage Directive sets a voltage limit of 50 V. In the United States, there is no federal law but local regulations to follow. Still, there is a national code (NEC) that sets the standard; the limit for low voltage is 100 V. Voltage limit may vary from region to region, but basically requirements on skill and equipment increase with voltage level. If going above the local voltage limit, there are more strict requirements on the design of cables, terminals and internal components. As these technical requirements have a long history, in most cases there is no doubt about how to imply them and how to build a device with a safe electrical functionality. In most areas and in general terms, compliance with the regulations is shown by an approval mark.

However, electrical safety is also about installation work on site. What are you (as a measurement technician or engineer) allowed to do? If the voltage is above the voltage limit, discussed above, basically you cannot do anything but replacement of components such as fuses and plugs. But again, regulations are local and vary, so you must first check what applies in your region! In addition to national law stating basic rules, in most factories there also local guidelines and 'licenses' for various works. The general advice is to check with your colleagues, be careful and always double check so that power is off before doing any work in electrical systems. If you anyway decide to do a modification or a change in an internal component, you should also remember that you have to take the responsibility of future safety in that specific device. The manufacturer's warranty (and/or safety marking) might now be invalid. Before doing any repair work that includes replacement of components, be sure to check with the manufacturer that you are allowed to do so. After you finished your work, when leaving an instrument connected to mains, always check that all cables are correctly installed and secured. Also check that cable glands, covers and lids are in position and without damage.

IP-Rated Enclosures

An important part of the electrical safety related to reliability and good functionality of a measuring instrument is the protection from dust and water. How 'tight' the enclosure is can be described in different ways. One is to use the 'IP-class'. The IP system code originates from the international standard IEC 60529 (Table 9.1). National Electrical Manufacturers Association (NEMA) offers a similar standard. NEMA Standards Publication 250-2003 divides enclosures into different classes depending on application (Table 9.2).

Electrical Disturbances (EMC)

Measuring instruments often have sensors, electrodes, cables and other components where signals can be 'exposed' to external signal sources (like mobile phones). If signal levels are small, it can be relatively easy to get the disturbance

TABLE 9.1

Enclosure IP-Classes

First Digit (Intrusion)	Second Digit (Moisture)
0 – No protection	0 – No protection
1 – Protection from a large part of the body such as a hand (but no protection from deliberate access); from solid objects greater than 50 mm in diameter	1 – Protection against condensation
2 – Protection against fingers or other object not greater than 80 mm in length and 12 mm in diameter	2 – Protection against water droplets deflected up to 15° from vertical
3 – Protection from entry by tools, wires and so on, with a diameter of 2,5 mm or more	3 – Protected against spray up to 60° from vertical
4 – Protection against solid bodies larger than 1 mm (e.g. fine tools/small)	4 – Protected against water spray from all directions
5 – Protected against dust that may harm equipment	5 – Protection against low-pressure water jets (all directions)
6 – Totally dust tight	6 – Protection against string water jets and waves
	7 – Protected against temporary immersion
	8 – Protected against prolonged effects of immersion under pressure

TABLE 9.2

Enclosure NEMA-Classes

NEMA Type	Situation	Possible IEC Equivalent
1	Indoor/general	10
2	Indoor/dripping water, falling dust	11
3	Outdoor/rain, snow, windblown dust	54
4, 4X	Hose-directed water, corrosion (X)	56
5	Indoor/angled dripping water, settling dust	52
6	Submersion	67
7	Hazardous area	
8	Hazardous area	
9	Hazardous area	
12	Indoor/dripping, dust	52
13	Indoor/oil, dust	54

imposed on the signal, for example, control problems or faulty totalisers as result. Therefore it is very important to be careful when connecting sensors and signal cables. Always check installation manuals and make sure that provided shields are correctly connected. If no instructions are available, it is usually good practice to use a shielded, twisted pair cable. Remember that even if you can verify the connection by checking the function of the system, the electromagnetic compatibility (EMC) protection is more difficult to check as it will be useful only when a disturbance occurs.

The term EMC is often used. There are two limits to observe, one for emission and the other for immunity. The idea is that no devices shall send out electromagnetic signals above a specific limit, and that the device shall not be disturbed by electromagnetic signals under a specific limit. However, for some systems, it might be that these general limits are not good enough. As an example, during a service job if you call someone to get assistance, standing in front of an electrical cabinet with open doors, you will probably expose the electrical equipment inside this cabinet with EMC levels above standardised limits.

Disturbances can be distributed basically in three ways: through the air, through cables or by contact. High currents in a cable will result in magnetic fields that can induce disturbances in nearby signal cables. Power supply units and power controllers (switching mode), especially old ones, will create disturbances on the power grid. Motors and certain types of lights will create a phase shift in the power supply, and this can also result in disturbances.

For very small and sensitive signals, there are professional cables and devices available. One example is shields made of mu-metal (or μ-metal). Mu-metal contains a lot of nickel, which is efficient to protect from magnetism. Another example is a boot-strap system, where electronic amplifiers will feed the shields with signal levels close to the 'real signal' resulting in a very good protection.

Pressure Safety

This section is about material strength and perhaps this is not normally covered by people working with instrumentation and automation. But as instruments also need to work under pressure, and as it is about health, safety and protection of assets, a short section about this is included. Basically, the media (liquid or gas), the size, operating temperature and pressure together set the design requirement. The same combination also sets required quality assurance arrangements. Compliance with the requirements is in many regions shown by the manufacturer by an approval mark, but the responsibility will be with the process owner for the safe management, operation and maintenance of the equipment.

At higher temperatures, the strength of most materials will be reduced. This means that the maximum pressure allowed in any given component will be lower with higher temperatures. In carbon steel, as an example, maximum pressure will drop with around 1 bar if temperature goes up from 150 to 200 °C.

Other materials have other relations, and at high pressures also steel has another relation. Therefore, always check both design temperature and pressure before selecting any components. Most components used in pipe lines are rated with a 'nominal' pressure. Observe that this nominal pressure is 'not' equal to the maximum operating pressure at maximum operating temperature for that specific device.

Material strength, as well as many other aspects, is closely connected to general quality assurance systems (such as ISO 9001). The basic idea of most of these systems is to divide the assurance into two parts. The first part is to describe 'how' the company normally works, and what material is delivered. The second part is to describe 'what' is delivered and to certify that the material was produced according to normal practices and routines. This means that there are often two parts of the quality documents to study: first procedures and secondly confirmations and certificates related to the procedures. Quality is a tricky concept, as most people translate it to 'good' quality. This is however not the case, basically a quality assurance system will not say anything about good or bad quality, but instead it will ensure that products have a 'constant' quality and that products are made according to what has been 'promised'.

Material Properties

As a mechanical strength calculation always includes raw data for the specific material in use, there is in some cases a need to verify that each device is made of the correct material. This information can be found in a material certificate. Standardised formats vary with regions, but a common type is in accordance with EN 10204. Mainly three different versions are used, and they declare that metallic products are in compliance with given requirements (type 2.1) and supplied test results (types 3.1 and 3.2). Types 2.1 and 3.1 are issued by the manufacturer, and type 3.2 is confirmed by an external, independent organisation.

Corrosion

Corrosion is the gradual deterioration of a material due to its environment. Iron, for example, will become 'rusty' in the presence of water and oxygen from the air due to the formation of iron oxide. If a measuring instrument starts to corrode, this will most likely affect both pressure safety and measurement uncertainty, and it is therefore important to always select wetted parts (parts in contact with the fluid) that are resistant to that specific fluid. For measuring instruments, corrosion is only about pressure safety, as material flexibility and other properties might also influence the performance. For example, the material in a pressure sensor membrane, the electrodes of an inductive flow meter or the tubes in a Coriolis flow meter must be selected with respect to what chemicals are used in the process.

Difficult Applications

There are a few commonly known fluids that are difficult to handle. Observe that a material used in a measuring instrument may require a higher level of chemical resistance compared to other components, as an instrument may also need to maintain properties such as flexibility and similar. High-quality materials such as titanium and tantalum will in most cases be resistant to most chemicals. But, there are examples of the opposite, for example, in systems with pure methanol (>98%), titanium or tantalum should not be used. Another example is that titanium can react very fast with oxygen, and in systems with oxygen gas (>35%) other materials must be used to prevent from explosion. In Appendix, you can find further tables with various fluids and compatible materials.

Hazardous Areas

For detailed information, study more about explosion safety in local regulations and/or specific courses on the subject. All legal requirements must be observed.

In a measuring instrument, there are a few aspects related to explosion safety that must be observed. Whether these are applicable or not depends on where the instrument is installed and in what application it is used. If a device is situated near (or inside) tanks, pipes or other systems containing flammable materials, it must be designed in a safe way. This would also apply to devices used in a process that may produce potentially flammable or explosive atmospheres. A design that guarantees that there are no ignition sources exposed can prevent an explosion if the device is exposed to gases, dust or liquids. These design requirements mainly cover electrical parts, but in some cases mechanical parts can also be included. The requirements extend to operating and maintaining such devices in the explosive, or potentially explosive, areas.

The Fire Triangle

Three things are needed to create a fire: flammable material, oxygen and an ignition source. To create an explosion, also a containment, a limited space is needed. Explosion safety is about removing one or more of these. In some cases, even very low energy (in the order of millijoule) is enough to ignite a gas or dust mixture. This energy is easily produced in the spark inside an electrical switch. Sparks can also occur if, for example, the battery of your mobile phone is disconnected (if you drop it and it breaks). A spark can also be created when two metallic parts hit each other in a mechanical device. Also, static electricity can cause sparks, and a petroleum product flowing with high velocity in a rubber hose can even cause static electricity by itself.

Inside pipes, tanks and components, the risk is often smaller, as there is no oxygen. This means that the risk might be higher during maintenance work, for example, when a petrol tank is drained and emptied. The small volume of petrol that remains on the floor, or the vapour in the air, can easily be ignited with an electrical tool such as a cutting machine brought by service personnel. A special warning is issued for people working with ethanol and methanol, as these liquids will create a mixture almost perfect for ignition above an open surface at room temperature. Similarly, special care should be taken if working with hydrogen. This gas is very easy to ignite and it will often leak and pass seals and gaskets, and usually, it is more difficult to detect hydrogen compared to other explosive gases. In general terms, 'heavy' gases require more attention than others. A heavy gas has a density higher than air, and it will therefore flow downwards and can be collected in deep locations such as sewage systems and closed tanks and rooms (Tables 9.3 and 9.4).

Risk Assessment

If prevention of explosive atmospheres is not possible, investigate how to arrange protective actions, how to remove ignition sources (or limit power to a safe level) and/or how to isolate a possible explosion in containment (or install explosion venting devices).

Electrical Equipment

No matter where you work, there are national regulations for electrical devices used in areas where explosive gases (or dusts) can occur. To fulfil the requirements in detail, you must read and understand these, probably together with an expert in this area. Basic recommendations are similar in most countries and include (1) risk assessment, (2) removal of explosive materials, (3) removal of ignition sources and (4) finally suppression (to limit

TABLE 9.3

Gas and Dust Groups

Substance	EN/IEC Group	FM/UL Group
Methane	I	I
Propane	IIA	IIA
Ethylene	IIB	IIB
Hydrogen	IIC	IIB H2
Acetylene	–	IIC
Combustible particles	IIIA	IIIA
Non-conductive dust	IIIB	IIIB[a]
Conductive dust	IIIC	IIIC

[a] Including carbon dust.

TABLE 9.4

Explosion Protection Standards

Type	Zone	Code	EN and UL	FM/ISA
Liquid immersion	1,2	o	60079-6	3600/12.16.01
Pressurised enclosure	1,2	p	60079-2	3620
Powder filling	1,2	q	60079-5	3600/12.25.01
Flameproof enclosure	1,2	d	60079-1	3600/12.22.01
Increased safety	1,2	e	60079-7	3600
Intrinsic safety	0,1,2	ia	60079-11	3610
Intrinsic safety	1,2	ib	60079-11	3600/12.02.01
Non-incendive	2	n	60079-15	3611/12.12.02
Encapsulation	1,2	m	60079-18	3600/12.23.01
Dust ignition protection	1,2	t	60079-31	3616/61241-1
Restricted ventilation[a]		fr	EN13463-2	
Constructional safety[a]		c	EN13463-5	
Ignition sources[a]		b	EN13463-6	
Liquid immersion[a]		k	EN13463-8	

[a] For non-electrical equipment.

the effect of a possible explosion). The basic idea in most regulations is to first decide what risks there are in a specific area. These areas are called 'zones' and the risk is indicated by numbers. For example, in the European ATEX system, the risk of explosion is highest in zone 0, and a typical example of a zone 0 area is inside a fuel storage tank. Local authorities will usually, in cooperation with the operator, set up where different zones are located. Gas types and temperature classes add more information about the specific application and what the risks are with specific substances.

Design Codes

Very often design codes are used to classify components to be used in ex-zone (Table 9.4). The class will provide, for example, tightness and robustness of an enclosure. In other systems, there are electrical 'fuses' that will protect the circuit. Two classes are commonly available in measurement circuits, 'ia' or 'ib'. 'i' stands for intrinsically safe and it basically means that even if two wires are touching each other, at a short circuit, the energy in the spark that will occur is not dangerous and is too small to ignite a possible gas or dust. This special fuse is called 'barrier' ('ex-barrier' or 'zener-barrier'). A normal fuse will protect from high current; this barrier will stop both high current and high voltage (thus limiting the energy). Type 'ia' has a higher safety compared with type 'ib'.

Other protection designs have the purpose to enclose and separate electrical components from possible flammable gas (or dust). Gas tight enclosures are marked with 'd' (or 'e' which in practice is similar). If the electric compartment is filled with some material, no gas can enter and these designs are marked with 'm', 'q' or 'o' depending on filler material. Air pressures slightly higher than the surrounding atmosphere will prevent gas from entering. This design is marked with 'p' and can be used for large constructions, such as a complete control room. Type 'n' (non-incendive) means that there can be no sparks under normal operation. This design is similar to 'i', but the safety level is lower for an 'n' device compared with an 'i' design because the 'i' design is evaluated to be safe also under fault conditions. These markings may be different depending on in which region you work and what standards are used, but basic ideas are very similar worldwide.

To summarise, you need to first find out if there are any explosion risks in the area that you will install your new measuring instruments. If there are flammable dusts, gases, vapours or mosts you should search for the zone map and see (or select) what zone the instrument is working in. You also need to know how 'dangerous' or flammable your specific application is (gases are categorised in groups). After knowing this, you need to select equipment to fit in this zone. Remember that mechanical devices can also generate heat or sparks and, therefore, needs to be designed in a safe way.

Electrostatic Discharge

Earthing is essential to prevent the build-up of static electricity. Before filling flammable liquids or gas from any vessel to another vessel or system, a cable that sets the electrical potential of all components to the same level should be connected. This is standard equipment on, for example, tanker trucks, but can also improve safety when filling small containers manually. If you work with electronic components, you should protect them from static electricity. This is not primarily a measurement problem, but as many meters include electronics it is good to know how to do this in a safe way. The low air humidity discharge level can reach over 30000 V.

Air humidity will affect the risk for static electricity. A free-falling jet of flammable liquid can at a certain distance and at low air humidity induce static electricity by itself, so special care should be taken when flammable fluids are poured, for example, when filling a tank (filling from bottom is always recommended).

Functional Safety

Accidents in chemical and process plants have led to injuries, death and damage to property and the environment, leading to new standards and routines for safe instrumentation and high-quality monitoring of potentially dangerous processes. This part is different from other safety issues discussed in this book, as now it is about functionality. For example, if a level meter does not indicate the correct level in a storage tank, this can be a dangerous situation. More recently, a series of standards have been published regarding functional safety. The International Electrotechnical Commission has published several standards regarding functional safety. IEC 61508 describes general requirements, and IEC 61511 has more details on instrumentation and measurement systems for process industries. There are also other standards in this family, describing special demands in various industries. Basically, these standards are designed around three facts: (1) the risk can never be zero, (2) non-tolerable risks must be reduced and (3) risks must be considered when designing a process. In many areas, similar approaches were also used before the implementation of these standards, for example, in nuclear power plants. The key principle for these standards is the safety lifecycle; applying safety throughout the entire design, installation, operation process to reduce the likelihood of systematic errors.

Safety Integrity Level

Just like in Ex-zones, equipment used in dangerous processes can be grouped into classes. SIL stands for Safety Integrity Level, and the level number indicates the amount of risk reduction provided by a system. In a SIL 1 system, the risk of failure has been reduced at least with a factor of 10, in SIL 2 at least 100, in SIL 3 at least 1000 and in a SIL 4 device the risk reduction factor is over 10 000. Of course, it is not easy to define either risks or what is judged as a 'fault', and to get a certification for a SIL application is a complicated and often time-consuming process. *All* devices incorporated in a safety system (not only a single component, e.g., a measuring sensor) need to be included in the safety assessment. To do this assessment, specific and statistical data are needed for each component in use. Obtaining these data can often be the most difficult part of the assessment. It is vitally important that the data used can be reference and validated, and that any caveats on the use of the data have been considered. For example, the failure rate of a component may be given that is only valid within certain temperature ranges. The SIL concept also applies across the entire lifecycle of the system development, and it is therefore also dependent on how a system is maintained and operated.

Traditional Concepts to Improve Safety

Analysis

To perform a risk analysis is essential, whichever standard is selected. There are several ways to do this analysis and some assistance may be required from other engineering disciplines and specialists. One method is called 'what if technique' and this is commonly used, easy to understand and built on practical experience. Among other methods, you will find HAZOP (hazard and operability study), FTA (fault tree analysis) and FMEA (failure modes and effects analysis). Another principle is ALARA (or ALARP), which stands for 'as low as reasonably achievable' (or as low as reasonably practicable). These principles are based on best practice judgement, and that the residual risk shall be as low as reasonably achievable.

Redundancy '2 out of 3'/'2 out of 4'

By connecting more than one sensor to measure the same parameter, safety can be improved; partly because if one sensor has failed there are others that remain in operation. Also precision can be improved, under special conditions, if the average signal from several sensors is used (compared to using only one sensor). You can read more about the precision aspect in the chapter on uncertainty. Normally, the purpose of using several sensors is safety. How these sensors shall be connected depends on what is considered to be 'safe'. In a '2 out of 4' configuration, it is possible to remove one sensor for service, calibration or similar without any impact on the safety. In addition to this, safety can be improved even more by the 'diversification' concept. The difference to the normal '2 out of 3' concept is that here sensors of different brands, working principles or similar are used. The idea is that if a process disturbance occurs, this will affect sensors of different principles differently, and therefore less susceptible to common cause failures.

Further Reading

Center for Chemical Process Safety. http://www.aiche.org/ccps/.
European Commission (ATEX directive). http://ec.europa.eu/.
FM Global© (explosionproof). http://www.fmapprovals.com/.
Health and Safety Executive (Hazardous Installations). http://www.hse.gov.uk/.
International Electrotechnical Commission (functional safety). http://www.iec.ch/.
Physikalisch-Technische Bundesanstalt (Explosionsschutz). http://www.ptb.de/.
Research Institutes of Sweden (process safety). http://www.rise.se/.
UL© (hazardous-locations). http://industries.ul.com/.

10

Calibration and Traceability

There are various standards that explain words and concept in this area, unfortunately not always in agreement. The vocabulary in this book is mainly based on ISO/IEC Guide 98/99.

How Exact Is Your Measurement?

It is important to remember that hardly any measurement is 'exact'. All measurements you perform have errors and faults. If you spend a fortune on a new instrument, it is still not exact, even if it presents data to you with many decimals!

Unfortunately, talking (and writing) about how correct a measurement is often causes confusion. For example, if you use the word 'exact', of course it matters what you mean by 'exact'. In theory, it means that it is absolutely true, no tolerances at all. It is not possible to state, for example, a temperature, pressure, length or mass without any tolerances. Even if tolerances are small, there will be tolerances! Here, we can also question the word 'small'. What might be a small tolerance in one process may be a very large tolerance in another. So, what is considered to be correct or not depends on how accurate the measurement is and how close to the true value you need to know this specific value. As long as the precision of the measurement is within the required tolerance, everything is fine. If not, there might be a costly and possibly dangerous problem. To find out whether this is the case or not, an uncertainty analysis might help. An uncertainty analysis is a good way to find, and reduce, errors and uncertainties, and you can read more about this in Chapter 11. First, we will discuss calibration and traceability as these are the basis for uncertainty.

Calibration

By definition, in measurement standards, 'calibration' is equal to 'comparison'. A calibration does not therefore automatically include any adjustment activities. This is a bit controversial and is often a cause for confusion in discussions and reports. Adjustments to a reading 'as close to correct as possible' can be performed at the same time as a calibration, but it is then something extra. Besides, if adjustments are performed, it might be a good idea to calibrate twice, before and after the adjustment. In many calibration reports, these two values are labelled with 'as found' and 'as left'.

The fact is that a calibrated instrument is not by some guarantee a 'good' instrument. However, it is documented how accurate it is.

To be able to decide how large the measuring error is, a better device or method (or at least a device with a known uncertainty) is needed. This is the reference. The precision, stability and confidence of the reference are very important. All calibrations must be recorded in calibration reports (or certificates). These documents shall include not only the result from the calibration (the deviation between the indication of the reference and the tested instrument, sometimes called device under test, or DUT) but also the method used, environmental conditions, the references used and the estimated uncertainty in the calibration. It must be possible to track the calibration of the used reference as well [it does not have to be stated in the report itself, but may be shown in a quality assurance (QA) system of the laboratory or similar]. As the reference used in the calibration also needs to be calibrated, you can follow this track of calibration all the way to the international standard for that specific unit. This is what traceability is about, and a calibration without traceability is of no value!

So, a calibration label on an instrument is in itself no quality mark for that instrument. The value lays in the document enclosed, where data and numbers can be used for confidence in future use of that specific instrument. The calibration report is also a good starting point for estimation of the measurement uncertainty.

If mistakes are made in a measurement, they might cause a problem to a product, batch or similar. If a mistake is made during calibration, the problem is often much worse as this will often cause problems over a longer period. If a mistake is made when calibrating a reference, problems can become huge. In this case, the error can be duplicated to many other instruments used in many applications over a long time. It is therefore crucial to be careful when performing calibrations, and it is important to select well-known sub-suppliers and partners if giving this job to others. One way to get some extra peace of mind can be to hire an accredited laboratory. An accreditation guarantees traceable references, validated methods and skilled personnel, and there are international standards stating how this shall be done. One popular standard for accredited calibration laboratories is ISO 17025, and in most countries, there are technical boards arranging assessments and certification of such laboratories.

Calibration is normally not a one-time activity. Instead, the calibration must be repeated at regular intervals. A tricky question is how long these intervals should be. There is no easy answer to this, even if very many people use 1 year by default. There is no technical reason for this specific period, but

TABLE 10.1

General Check and Recalibration Plan

	Check	Test	Calibration
Number of data points	One	A few points	Complete range
Interval	Before use	Weeks/months	Months/years
Performed by	User	Maintenance dept.	Calibration dept.
Documentation	None	Journal/note	Report

it may be convenient and practical as systems can easily be set up to meet the 1-year repeat. For a more technical approach, there are basically two things to consider: the drift of the instrument and the severity in the risk of having errors in the measurement. If possible, a good way is to find individual calibration intervals. When the instrument is new, this can be performed as follows: After 6 months (or at any suitable time), the instrument is recalibrated. If results are equal to those when the instrument was new, the interval can be extended and next calibration can be performed in a year. However, if results have changed and the instrument seems to be drifting away or in any other way being influenced by the application and the environment, the next calibration should be performed sooner and within a shorter interval. Using this method, a 'history' will be created for each measuring instrument, and this is a good base when deciding recalibration periods (Table 10.1).

Calibration Report

After performing a calibration, a report or certificate should be produced. First, the calibrated instrument must be identified: state manufacturer, type, size, version serial number and tag number if there is one. Then, make notes not only of measured results (readings of reference and DUT, including what signal outputs have been used) but also what equipment and methods you used and various conditions like pressure, temperature, ambient temperature, location, time and date, operator and everything that can be of use later. Do not forget to include information on all reference instruments used, preferably including information (document ID) on their calibrations. Finally, state the estimated uncertainty.

Traceability

As mentioned several times in this book, traceability is a must for all measurements and calibrations. In each calibration report, the traceability should be stated in the form of references to calibration reports for those instruments used to produce 'true' values in the report. This can result in long chains of references, and if done correctly, these chains will connect each individual instrument to international standards. As you probably can imagine, this can result in quite a complex system that is not at all easy to verify and maintain. For this reason, there are local, national and international organisations to support you. Some organisations do the calibration work themselves (like national laboratories), and others only observe, issue accreditations and assess calibration laboratories. All this work originates at BIPM (International Bureau of Weights and Measures) in Paris, where the world standards for many measurements and units can be found. In the United States, it is common to make a statement that a measurement is 'traceable to NIST'. As NIST (National Institute of Standards and Technology) is traceable to BIPM, this has then basically the same meaning.

Concepts

Explanations, assistance and guidance can be found in both national and international standards (like ISO Guide 99). Definitions and expressions are sometimes used with different meanings in the 'real world'. For example, there are several books and dictionaries explaining the meaning of the word 'calibration' as an 'action to improve performance'. In addition, this is not correct if referring to measurement standards. Also, other words and concepts might have several meanings, so be careful. What does the information stated on the back of a data sheet really mean? What does a manufacturer mean with 'accuracy' or 'response time'? Even if there are standards to explain this, one cannot be sure that the writer did use these standards.

Error

According to the standards, measuring error is something that is known: a number, often in percent, telling us how far away from the true value a measurement is. That means the error has a sign: positive or negative. If the error really is known, it is possible to correct for the error and correction is equal to the error with the opposite sign. In practice, however, it is quite common that the error is *not* known. For example, most data sheets state an error limit; a product is guaranteed to stay inside a specific error limit (e.g. ±1%). Then, in this case you do not know the error in detail and you cannot compensate for it, but you know it is not larger than 1%. One common way is to express the error as percent of reading, meaning the specific measured value where the error was observed. Another way is to express the error as percent of full scale, meaning the largest possible value of the calibrated instrument.

Let us look at an example. A pressure sensor with a measuring range of 0–10 bar is calibrated. When 5 bar is applied, the sensor indicates 4,9 bar. The error is then −0,1 bar. In percent, the error can be stated as −2% of reading or −1% of range.

Repeatability

Repeatability is a quite good indicator of instrument quality. If an instrument has a good repeatability, it can reproduce its measurement and that means you can trust you will get the same reading many times if measuring on the same product. However, a good repeatability is not equal to a correct measurement. In most cases, an instrument with good repeatability can be adjusted to also give a correct answer. A typical example of such a case is an instrument of good quality that was never calibrated.

Accuracy

Another commonly used word is accuracy. This could be explained as measuring error and repeatability combined.

Uncertainty

Measurement uncertainty is explained in detail in Chapter 11, but a short description would be the final error of a measurement result, a grey zone around a value that (in contradiction to the error) is never known.

Intercomparison

With a close connection to calibration activities and accreditation, an intercomparison is a commonly used procedure to ensure that a calibration laboratory delivers good quality results. Sometimes, this activity is called a round robin test. The idea is to take a good and repeatable instrument and send this to a group of laboratories. They should all perform the same calibration, and later all results are compared by the organiser. If, for example, there are eight participating laboratories in a pressure intercomparison and seven of them report a measuring error of about 0,2 kPa and one reports 0,8 kPa, this probably indicates that lab no. 8 has a problem. All results should be compared together with stated uncertainties, and if one laboratory is still 'outside', something may be wrong. At least, it is a reason for further investigation. If two similar instruments, measuring identical data, are used for the intercomparison, this will allow even deeper analysis of the laboratory equipment. One way to use double measurements, for example, two pressure sensors in parallel or two flow meters in series, is a cross-correlation analysis (or a Youden plot), a very informative way to present the laboratory test results. The idea is that both instruments should be calibrated together with several repetitions.

If both signals follow each other such that in calibration no. 1 both are low, in calibration 2 both are high, in calibration 3 both are in between and so on, there is a correlation. This calls for an investigation of the calibration equipment. If all results are random (which is preferred), the plotted calibration results will form a circle in the diagram. A strong correlation will result in a 45-degree line across the diagram.

Conformity Assessment

This is a process, or action, often performed by authorities or independent organisations. However, it can also be an internal decision, based on one or several measurements. The purpose is to find out if a product (or action) is within stated limits. One example is police performing a speed control on the highway. They will then measure the speed of the vehicles passing, and if someone drives too fast, possibly the driver must pay a fee. When deciding whether the speed is too high, the police must do a conformity assessment. This should then include not only the measurement itself, but also the measurement uncertainty. Normal procedure is to decrease the measured speed with the uncertainty to make a fair judgement.

Risk

Related to the conformity assessment, there are three ways to handle the uncertainty when making a decision based on a measurement. The 'seller' can take the risk, the 'buyer' can take the risk or they both can share the risk. In the example with the police, he or she will take the risk and reduce his or her measurement with the uncertainty. In a safety application, probably the client should take the risk and increase his or her measurement, for example, temperature, to stay within limits and guarantee a safe operation. In a shop, buying vegetables for example and paying for the weight, the buyer and seller will share the risk. This results in a payment corresponding to an uncorrected mass.

Signal Loops

If the sensor or instrument you are calibrating is a part of a system or measuring loop, the complete system or loop of course must be calibrated. For practical reasons, this is usually made in more than one step. We can take a temperature sensor as an example:

First, the sensor is calibrated with a block calibrator. This is usually done on site but with the sensor removed from the process. As the transmitter is integrated inside the sensor, this component was included in the calibration.

Next step is to calibrate the signal transmission to the control system. This is performed with a signal generator replacing the temperature transmitter. The output of the control system (or the reading on the display) is then compared with the simulated signal. Now, the complete system is calibrated and the total uncertainty can be calculated based on the two results combined.

Further Reading

ISO/IEC Guide 99 – International vocabulary of metrology (VIM).
IEC 61298 – Process measurement and control devices.

11

Measurement Uncertainty

Better roughly right than precisely wrong

Carveth Read/John Maynard Keynes

No measurements are correct. Luckily, in most cases, the errors are so small that we can forget about them. We can live with small errors; small deviations due to mistakes in measurements will not likely affect production, quality or similar. However, sometimes errors are simply too large to ignore. In such cases, we need to act to increase the precision by changing instruments or by modifications in the process. The problem is that it is not easy to know when errors are small or large. To get a value of the measurement error, you need to perform a calibration. There are also methods, some using statistics and mathematics (trying to balance energy input and output), to find out possible errors, but by far calibration is the most common procedure. In any case, when taking decisions based on measurement data, it is essential to know how large an error the measured data has. This chapter will give an introduction to uncertainty calculation – a standardised way to estimate how good a measured value is.

Numbers

As an example, a measurement result can be written as 18,5. In many cases, it would be more appropriate to also state the uncertainty and write 18,5 ± 0,2 (as an example). The meaning of this is that the true value lies between 18,3 and 18,7. If the uncertainty was instead ±0,02, it would be more appropriate to write the measured value with one more decimal place, 18,50. On the contrary, with an uncertainty of ±2, it is not useful to write 18,5. To write 'too many' decimals is truly misleading! Normally, the number of decimals should be related to the uncertainty, but this may not always be the case!

So please observe! An instrument with a display that shows three decimal places (0,000) does *not* by definition have a smaller measurement error than one with two decimal places (0,00).

Also the resolution of a display is not always the same as the last digit. There can be two digits (or any size) between the smallest 'step', that is, 0,00 – 0,02 – 0,04 – 0,06 and so on.

Conformity Assessment

It is not unusual that measurement results are used for checking or verification purposes. It could be some form of law enforcement (like police doing speed control on the highway), but it could just as well be an internal quality check to verify some aspects of a product. It is important to adjust for the measuring uncertainty when performing such activities, conformity assessments.

Basically, the uncertainty can be handled in three different ways. Either it is the 'supplier' or the 'customer' who takes the risk or else, the risk is shared between both of them. Risk here means the uncertainty. Let us take an example: you buy 1,55 kg apples and the weighing scale in the shop has an uncertainty of 0,05 kg. If you take the risk, you will pay for 1,60 kg (measured value *plus* uncertainty). If the shop takes the risk, you will pay for 1,50 kg (measured value *minus* uncertainty). However, in a shop, shared risk is what is used, and therefore, you pay for 1,55 kg. In addition, there are rules about maximum uncertainty for weighing scales in shops. When the police check speed, they take the risk, normally by reducing the uncertainty from the measured value before writing a speed ticket. When monitoring pollution limits, maximum power and similar, it is in most cases the customer (i.e. a power plant) who takes the risk, so that values are guaranteed to be below limits. This has the effect that a reduced uncertainty may allow for a higher production, since measured values can be closer to the limits.

Comparisons

A comparison of two measurements is in some ways like the conformity assessment described above. If, for example, you put two temperature sensors in the same pot of boiling water, you will probably get two temperature readings. The difference will depend on two things, how even the water temperature is (the temperature distribution) and how large uncertainty each temperature reading has. Without knowing the uncertainty of each sensor reading, you cannot tell if there really is a temperature difference in the pot or not. Only if the difference you see is larger than the uncertainty, you know that there is a real difference in temperature at different positions. To be precise, this should be treated as a statistical event, with two normal distributions. If the two areas overlap, both readings are in consensus; if there is no overlap, there is a real difference.

If you can tell from other information that there cannot be any differences (e.g. two flow meters after each other in the same pipe, measuring the same flow), the method described above can be used to judge if both measuring instruments are good. If there is no overlap, either one instrument is malfunctioning or the stated uncertainty is too small. In Chapter 10, you can also read about *intercomparison*, a quality assurance activity performed in laboratories.

Process Control

Compared to conformity assessment, uncertainty has similar effects in process control; however, in this case, large uncertainties may result in the need for larger 'production safety margins'. For example, in a heat treatment process, it might be needed to heat the product to a higher temperature than is required because of uncertainties in temperature control. To be sure, we heat to 75 degrees even if only 70 is needed. This of course costs both time and energy! In other words, a reduced measurement uncertainty can lead to a more efficient production and lower energy costs.

Uncertainty

As you know, not even calibrated instruments are exact; there are still error sources that cannot be avoided. Additional errors caused by drift over time, installation effects, signal transfer, recalculations and similar are always present. If errors are known, we can adjust for them, but in most cases we do not know them in detail. But maybe we can estimate error limits, or 'grey zones', in such a way that the error most likely will be inside these limits. Several grey zones (or uncertainty contributions) can be combined with a total zone inside which the final measuring result will stay. It is this total grey zone we call measurement uncertainty. There are standardised rules on how to combine various contributions into a total estimated uncertainty, telling how to add random meter errors to systematic errors. In practice, this often means you need to add apples to pears (e.g. an error in °C with an error in kPa). The setup of a measuring system normally calls for a sensitivity coefficient. This coefficient indicates how 'important' each measured data is for the result. For example, an error of 1% in temperature will not always cause an error of 1% in the total result. ISO Guide 98 indicates how to do this in detail.

There is a difference between the concepts of 'error' and 'uncertainty'. Simplified, you could say that error is something found in an instrument, and uncertainty is a result. Errors are sometimes known (e.g. after a calibration), but uncertainty is never known (it is an estimated zone). Precise definitions of these concepts are available in ISO Guide 98 and 99. But be aware that the standard definitions are not always used 'in practice' and that there is a need to discuss and explain when talking about these things in detail.

Random or Systematic

The uncertainty consists of two parts: one systematic and one random. The systematic part is the measurement error in your instrument, a value that is fairly constant – no matter how many times you repeat a measurement, you will always get the same error. Differences due to variations in ambient temperature, position/alignment, power supply, humidity, personal skill and similar will cause different results when repeating the measurement. This difference is the random uncertainty.

If looking at statistics, we can try to describe how results are varying if repeating a measurement several times. Totally, random data will result in a normal distribution. If plotting all results in a picture, the value seen most often is the average, and this becomes the centre. On both sides, there are values more or less far away from the average. The farther from the centre (and the average) you go, the fewer data points you will find. The distance to the average can be expressed with the confidence level. A confidence level of 95% is commonly used for measuring results, meaning that within these limits you will find 95% of all data. A confidence level of 95% is also known as 2σ (two sigma).

When repeating measurements, not all actions result in a normal distribution. Two common examples are when reading a digital display and when reading a mechanical ruler. The digital display cannot indicate anything in between two digits. If the last digit (it doesn't matter if it is a decimal or not), for example, shows 7, you do not know if it is close to 6,6, close to 7,0 or close to 7,4. The true value can be anywhere in this span, and this distribution is said to be 'square'. When reading an old thermometer, a ruler or something else with a mechanical scale, you can estimate values also in between numbers. This capability causes the distribution to be triangular. These three distributions, normal (1), triangular (2) and square (3), are what we use for uncertainty calculations, and depending on what distribution we use, the value is weighted more or less when we add all uncertainty contributions and calculate the total uncertainty. Weighting factors stated in Figure 11.1 are from ISO Guide 98.

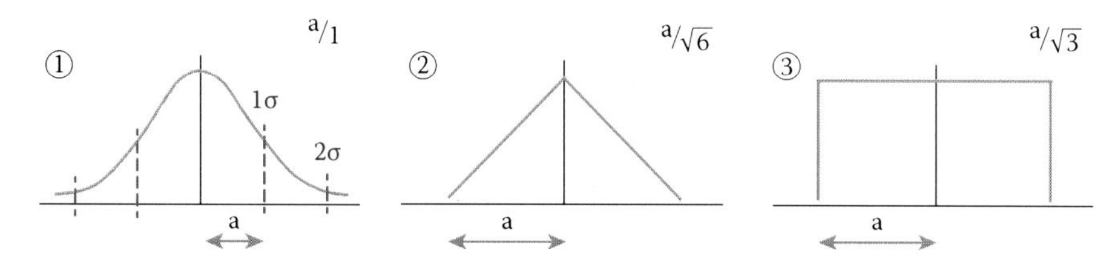

FIGURE 11.1
Statistical distributions.

Combining Results

Also the number of data sources will affect the distribution; using many sources will result in a normal distribution. You can test this yourself by a simple experiment. Roll a dice 50 times and note how many times you get 1, 2, 3, 4, 5 and 6. Make a diagram as in Figure 11.1 and insert the numbers (one bar for each result). Probably, you now have something similar to the illustration to the right in Figure 11.1. Now repeat this using two dices. What distribution do you get now? And what happens if you use three dices?

Uncertainty Budget

The total uncertainty of a measurement system (or inside an instrument with several sensors) can be estimated by doing an uncertainty budget. Furthermore, such a budget can also be a tool to assist in the improvement of a system as the size of each contribution will be clearly visible. The word 'budget' also indicates that this is still an estimate – perhaps one can claim that it is guessing, *but guessing with qualification and confidence.*

To sum all contributions, there are guidelines. The first step is to use experience, skill and imagination to find out all parameters that can contribute to the uncertainty. The second step is to find out how large each contribution is. Using a combination of data sheets, calibration reports, articles and experience, a value for each part can be set. The third step is about statistics and how to adjust the values according to their importance. All additions should be made at a confidence level of one sigma, so if values are stated at two sigma (which they normally are) they should be divided by 2. If distributions are square or triangular, the divisors indicated in Figure 11.1 can be used. Step number 4 is then to adjust for sensitivity, so that each part will influence the final measurement result like it would during operation.

Fifth step is to add everything, and this is done by quadratic addition.

As most standards require us to declare the uncertainty with a confidence of 95%, we finally need to adjust the result to two sigma by multiplication by 2.

The method described above has been in use for many years, but still it is not the only way to express uncertainty. You could, for example, just add all contributions, resulting in a slightly higher uncertainty. Anyway, in most cases, the important part is how you estimate each input. And when you can identify the highest contributor to the total uncertainty, you also know where an improvement will have the highest impact.

In some cases, it is more 'understandable' to use percent when indicating the uncertainty. For example, a flow rate could be expressed 500 L/s ± 5 L/s or 500 L/s ± 1%. To use percent makes it easier to judge if the measurement data are good or not, at least for an experienced technician. Uncertainties expressed in percent can be used for all types of data, but they are normally not used for temperature readings or other data that can be close to zero.

Both NIST in USA and NPL in the United Kingdom have published guides for uncertainty estimation in the English language.

Example

We want to measure level in a tank by means of a pressure sensor placed inside the tank, at the bottom. When installing the sensor, the tank is already in operation and filled with a liquid. Therefore, the sensor is installed by just dropping it from above, letting it sink and get to rest in any position at the bottom. So, how can we calculate the uncertainty of the level reading in this application? Firstly, we must understand the measurement and calculations used. Secondly, we must figure out what factors and issues there are that can influence our results and find reasonable figures to insert for each contributing factor. Finally, all these contributions should be combined to a total uncertainty.

What can influence the result of this measurement? Well, of course, the error of the sensor itself is a starting point. In this case, this instrument was not calibrated, so we only have the data sheet (saying that maximum error is 100 Pa). Next step is to have a look at the installation. As we just dropped the device from the surface, we do not know the position. Looking at a drawing, we see that the sensor has four sides, but it is not square. One side is 40 mm wide and the other 60 mm wide. This gives us the information that the centre of the device might be either 20 or 30 mm from the tank bottom. The tank is 2 m tall, but normally maximum level is 1 m and we should estimate the uncertainty at the normal level. Because of the dimensions of the sensor, we need to reduce the range with either 20 or 30 mm. We select to set the range to 975 mm.

As we use a pressure sensor to measure level, we need to calculate the measured value to get the level we want. In this calculation, we use measured pressure as input, and then, we need to adjust for the density of the liquid in the tank. We can calculate density from a table if we know the temperature. For the uncertainty calculation, we also need temperature limits. What are the temperatures of the coldest and warmest liquid that can be stored in this tank? If we know this, we can check what densities these temperatures correspond to and in that way find the uncertainty contribution.

TABLE 11.1

Uncertainty Example

Contribution	Estimated Value	Probability	Standard Contribution	Sensitivity Coeff.	Result (%)	Result^2
Sensor (data sheet)	±100 Pa	Square	/1,73	1/9000	0,64	0,41
Reading (display res. /2)	±1 Pa	Square	/1,73	1/9000	0,01	0,00
Position of sensor (height above tank bottom)	±5 mm	Square	/1,73	10/9000	0,32	0,10
Temperature measurement	±2°C	Square	/1,73	0,001	0,12	0,01
Temperature effect on level sensor	±2 Pa	Normal (2σ)	/2	1/9000	0,01	0,00
Density (level calculation)	±0,4%	Normal (2σ)	/2	1	0,2	0,04
Drift (1 year)	±150 Pa	Normal (2σ)	/2	1/9000	0,83	0,69
			√ Sum			1,12
		Sum adjusted to 95% confidence level (= sum × 2)				2,24

Can the temperature have any other influence? When checking the data sheet of the sensor, we find a temperature statement for accuracy dependency. This must also be added. What other effects can we find? If the tank is not completely open at top, there could be an overpressure that would affect the measurement. In this case, the tank has a wide opening so we do not have to consider this. Are there any additional errors in signal transmission from the meter? Not in this case. From the data sheet, we can see that the display resolution is not 'real', and the resolution is instead equal to a value corresponding to two last digits of the display. Finally, we need to estimate the drift in time. As we have similar instruments installed at another plant, we can check the history of these. The average difference between calibrations gives us good information about drift. If such history is not available, maybe a manufacturer's statement can be found. Now we think we found all uncertainty contributions and these can then be added as in Table 11.1. To make everything easy to compare, we can calculate the result in percent, where 100% is equal to 1000 mm which (with a liquid density of 900 kg/m^3) corresponds to 9000 Pa. To use 975 mm instead of 1000 mm would also be possible. The actual density is used from measured temperature and a table. Temperature is measured with a maximum error of 2 °C, and liquid density will change approximately 0,1% per degree Celsius. Also because of impurities and variation in concentration, nominal density value will change from day to day. This variation is estimated to be 0,4%.

The uncertainty is estimated to be 2,2%, which means that measured level can be expected within ±2,2% of the 'true' level.

Improvements

An uncertainty budget can also assist us to find best actions for improvements. This calculation indicates that the two largest contributions are the initial error (the specified accuracy) and the long-term stability (drift). If an initial calibration had been performed, and the reading was adjusted according to the measuring error found during this calibration, the value from the data sheet (±100 Pa) could be replaced by the calibration uncertainty. This would probably be a much smaller value. A more frequent re-calibration could reduce the long-time stability value (now stated as ±150 Pa/year) and thus could also reduce that contribution.

Further Reading

Bevington, P. R. 1969. *Data reduction and uncertainty analysis for the physical sciences.* McGraw-Hill.

ISO Guide 99 (GUM). http://www.bipm.org/en/publications/guides/#gum.

National Institute of Standards Technology. http://physics.nist.gov/cuu/Uncertainty/.

12

Foundations of Metrology

History

Measurements and traceable measurements have been performed for thousands of years. Among the oldest references we know today is the Egyptian Cubit (a length standard that had the same length as the arm of King Pharaoh). Cubits were used to measure buildings, lands and water levels. What legal requirements they may have had 4000 years ago is hard to tell, but regulations, including punishments for those who failed to follow them, did exist.

Natural objects, like seeds or body parts, remained the common way to establish measurement references for thousands of years. However, as body parts vary in size from person to person, it is also needed to select a reference person. Normally that would be the king, emperor or similar, but how then to organise international trade? With another king in a different country having, for example, longer feet, the 'real' length of something measured in feet (and the cost) could be a reason for big discussions. A different system was needed, and in the late 1800, a group of French scientists founded a new system: a system without kings and body parts. Instead, they decided to use our planet as reference. The system they invented was the metric system, well known today. One metre is the circumference of the world divided by 40 000 000. They measured this distance only partly, from Barcelona to Dunkerque, and calculated the rest with a very good result.

Today, the metric system is extended to many units known as Système International d'Unités (SI). SI units cover everything, but are not accepted everywhere. Units other than SI units are used in, for example, both the United States and the United Kingdom (sometimes called imperial units). Still, both these countries have adopted SI units.

Weights and Measures

When trading, for example, food and fuel, price is set by mass or volume. Therefore, weighing scales and measuring glasses are often used in shops and markets. For some salesmen, to deliver a small amount less than promised can be a temptation hard to resist. It is difficult for the customer to notice this, and even if it is observed, it is not easy to present any evidence. This can create endless discussions and disputes. To prevent people from cheating, already long time ago special officials were appointed to check measurement devices used in business. For example, in Greece an office was built already 1500 years ago – Tholos. Here, local business men had to get an approval of their measuring instruments. A seal was attached to the instrument to indicate to customers that they could feel secure. What we do in shops and petrol stations today is more or less identical with these old methods. Today, most countries also require a type approval (certification) for measurement instruments used in trade. In addition to maximum allowed errors, certification requirements include resistance to environmental disturbances such as heat, cold, humidity and electromagnetic interference. There should also be a clearly visible display and reasonable measures to protect against fraud. The details of these requirements vary from region to region. International recommendations are issued by Organisation International du Metrologie Legal (OIML). In Europe, common regulations are stated in the Measuring Instruments Directive (MID). In the United States, more information about regulations can be found at the National Institute of Standards and Technology (NIST), where Handbook no. 130 is a good starting point. See Chapter 13 for addresses and information regarding other countries.

To promote international trade, in the year 1999 several countries signed a mutual recognition agreement (MRA) related to measurements and accredited calibrations. If export trade (and invoice) is based on a measuring instrument having an accredited calibration, measured values shall be accepted by all countries that signed this MRA.

In most countries, today governmental offices are established to monitor the use of weighing and measurement equipment in use. The measurement itself can be called a fiscal measurement or, without having any governmental bodies included for inspection, custody transfer measurement. Governmental activities are often referred to as Weights and Measures acts. There are similar requirements for measuring equipment used for packaging. If you buy a bottle or package with a specified content, like 1 kg or 1 litre, there are requirements on the accuracy of these labels and the 'real' quantity of each package. These requirements have a statistical approach, and basically the mean value in a number of packages must not be too low.

SI Units

The Bureau International des Poids et Mesures (BIPM), established in 1875, works for a single system of measurements to be used throughout the world.

TABLE 12.1

Base Quantities and Base Units Used in the SI

Base Quantity	Symbol	Base Unit	Symbol
Length	l, h, r, x	Metre	m
Mass	m	Kilogram	kg
Time	t	Second	s
Electric current	I, i	Ampere	A
Thermodynamic temperature	T	Kelvin	K
Amount of substance	n	Mole	mol
Luminous intensity	lv	Candela	cd

TABLE 12.2

Derived Quantities and Units (Examples)

Quantity	Symbol	Unit	Symbol
Area	A	Square metre	m^2
Volume	V	Cubic metre	m^3
Speed, velocity	v	Metre per second	m/s
Acceleration	a	Metre per second squared	m/s^2
Density	ρ	Kilogram per cubic metre	kg/m^3
Frequency		Hertz	Hz
Force		Newton	N
Pressure		Pascal	Pa
Energy		Joule	J
Electric potential		Volt	V
Electric resistance		Ohm	Ω
Electric conductance		Siemens	S

TABLE 12.3

Examples of Commonly Used Non-SI Units

Quantity	Unit	Symbol
Time	Minute	min (= 60 s)
Time	Hour	h (= 3600 s)
Volume	Litre	L or l (= 1 dm^3)
Pressure	Bar	bar (= 100 kPa)
Pressure	Millimetre of mercury	mmHg (= 133 Pa)
Length	Nautical mile	M (= 1852 m)

TABLE 12.4

SI Prefixes

Factor	Prefix	Symbol
10^{24}	yotta	Y
10^{21}	zetta	Z
10^{18}	exa	E
10^{15}	peta	P
10^{12}	tera	T
10^{9}	giga	G
10^{6}	mega	M
10^{3}	kilo	k
10^{2}	hecto	h
10^{1}	deka	da
10^{-1}	deci	d
10^{-2}	centi	c
10^{-3}	milli	m
10^{-6}	micro	µ
10^{-9}	nano	n
10^{-12}	pico	p
10^{-15}	femto	f
10^{-18}	atto	a
10^{-21}	zepto	z
10^{-24}	yocto	y

The foundation is the metre convention, but the system has developed so that it now includes seven base units. In 1960, the International System of Units (SI) was introduced. More information can be found at www.bipm.org (Tables 12.1–12.4).

Other Commonly Used Units

Mass

Dram (UK) = 1,77 g

Carat = 0,2 g

Ounce, oz (UK) = 28,3 g

Pound, lb (UK) = 453,59 g

Stone (UK) = 6,35 kg

Tonne, long ton (UK) = 1016 kg

Tonne, short ton (US) = 907 kg

Volume

Barrel (UK) = 164 litres

Barrel (US) = 115–119 litres

Barrel, petroleum (US) = 159 litres

Cubic foot = 28,3 litres

Cubic yard = 765 litres

Fluid ounce (fl oz) = 0,028 litres

Gallon (UK) = 4,55 litres

Gallon (US) = 3,79 litres

Length and Area

Acre (UK) = 4047 m^2

Foot (UK) = 30,48 cm

Inch (UK) = 2,54 cm

Line (UK) = 2 mm

Mile, statue (UK) = 1609 m

Yard (UK) = 91,44 cm

Force, Power and Energy

Pound force, lbf = 4,44822 N

Kilopond, kp = 9,80665 N

Horse power (metric) = 735 W

Horse power (UK) = 746 W

Kilocalories, kcal = 4,19 kJ

Kilowatt-hour, kWh = 3600 kJ

British thermal unit, Btu = 1055 J

13

World of Metrology

Most activities, methods and actions discussed in this book are the same, wherever in the world you work. However, details may vary and standards sometimes have different names. In most countries, there are authorities and organisations to support your activities. In this chapter, a few organisations working in the field of process measurement are listed. Unfortunately, this list will not cover all regions, but if contacting any of these organisations, they are probably able to guide you to the correct destination.

Organisations

There are three different types of international organisations dealing with process measurements:

1. General standardisation bodies (like ISO)
2. Communities (like NAMUR)
3. Umbrella organisations for legal metrology and traceability (like OIML)

Additionally, each country usually has a national metrology laboratory (Table 13.1). Below is a list of various organisations, including a short introduction and contact information. Often this is a good starting point for further contact and advice.

AFRIMETS

Intra-Africa Metrology System provides standards and support for member nations within the African continent. *www.afrimets.org*

ANSI

American National Standards Institute is a community for standardisation in the USA. Measurement and calculation standards are available, as well as 'test schemes', for example, to test and verify machines and systems. Pipes, fittings and flanges with measures standardised by ANSI are commonly used worldwide. *www.ansi.org*

API

The American Petroleum Institute is an important organisation in the oil and gas industry. Many standards related to measurements have been published and related information such as density tables can be found here. *www.api.org*

ASEAN

The Association of Southeast Asian Nations is a Pan Asian organisation for traceability and calibration. *www.asean.org*

ASME

The American Society of Mechanical Engineers is a community for standardisation in the USA. ASME is well known for standards regarding safety of mechanical equipment, particularly boilers and pressure vessels. Measurement and calculation standards are available, as well as 'test schemes' and similar, for example, to test and verify various machines and systems. *www.asme.org*

BIPM

The International Bureau of Weights and Measures is the home of the metric system, and a foundation for traceable measurements. Here, you can find information, recommendations and advice as well as contact information for all member states. *www.bipm.org*

BMTA

British Measurement and Testing Association deals with members' concerns regarding measurements. *www.bmta.co.uk*

CEESI

Colorado Engineering Experiment Station, Inc., is a U.S. calibration laboratory for flow meters, also offering a variety of useful publications and documents. *www.ceesi.com*

DIN

German national standards, in many cases transformed to European standards. *www.din.de*

EA – European Accreditation Council

Supervising local and national accreditation bodies with applied accreditation and assessment of accredited laboratories. *www.european-accreditation.org*

EASC

EuroAsian Interstate Council works with standardisation, metrology and certification. *www.easc.org.by*

EI – Evaluation International

A cooperation between France, the Netherlands and the United Kingdom. Members (companies) each year arrange evaluations of various measuring devices to assist in selection and purchasing of new instruments. *www.evaluation-international.com*

EN/CEN

The European Committee for Standardisation is the centre for standards in general, and for electrical devices in particular. *www.cen.eu*

EURAMET

The European Metrology Programme for Innovation and Research is synchronising parts of the European research work, mainly related to national metrology institutes. *www.euramet.org*

GCC–GSO

The GCC Standardization Organization (GSO) is a standards organisation for the member states of the Gulf Cooperation Council, and a portal for standards, including metrology and measurement, is available on their home page. *www.gso.org.sa*

GOST

The Federal Agency on Technical Regulating and Metrology of the Russian Federation, home of Russian standardisation and legal metrology, publishes standards, recommendations and legally binding documents. *www.gost.ru*

ILAC

The International Laboratory Accreditation Cooperation assists national accreditation bodies with applied accreditation and assessment of accredited laboratories. *www.ilac.org*

IMEKO

IMEKO has sub-divisions for various measurement areas, such as temperature (TEMPMEKO) and flow (FLOMEKO). New findings and research work are presented at annual meetings and conferences. *www.imeko.org*

ISO

Worldwide organisation for standards of all kinds. Several standards related to measurements are listed elsewhere in this book. Regarding flow meters, one important example is ISO 5167 dealing with orifice plates and Venturi tubes. *www.iso.org*

NAMUR

German/European community for various technical issues. Two examples in the field of measurements are signal levels and explosion proof design where NAMUR standards are commonly used. *www.namur.net*

NIST

The National Institute of Standards and Technology is the U.S. measurement centre, located just outside Washington, DC. NIST is the home of American measurement references and the expression 'traceable to NIST' is famous. *www.nist.gov*

OIML

The International Organization of Legal Metrology is a sister organisation to BIPM, but with a focus on legal metrology (mainly measurements involved in payments, taxes and similar processes). *www.oiml.org*

SIM

The Inter-American Metrology System covers measurement and metrology issues for the member nations of the Organization of American States (OAS). *www.sim-metrologia.org.br*

WELMEC

With members from most European legal metrology institutes, their focus is to create a uniform approach to both national and EU laws related to legal metrology. *www.welmec.org*

WGFF

The Working Group on Fluid Flow is a working group within BIPM. Within BIPM, there are working groups performing various tasks related to practical measurements and other actions to secure international traceability. One important task is to arrange international measurement comparisons ('Round Robin' calibrations).

TABLE 13.1

National Metrology Laboratories

Country/Region	National Metrology Lab	Internet
Austria	BEV	www.bev.gt.at
China	NIM	www.nim.ac.cn
China	SITIIAS	www.sitiias.cn
Czech Republic	CMI	www.cmi.cz
Finland	MIKES/TUKES	www.mikes.fi
France	LNE	www.lne.fr
Germany	PTB	www.ptb.de
Greece		www.eim.gr
Italy	Inrim	www.inrim.it
Japan	NMIJ	www.nmij.jp
Netherlands	VSL	www.vsl.nl
	NMI	www.nmi.nl
Norway	Justervesenet	www.justervesenet.no
Russia	VNIIM	www.vniim.ru
South Korea	KRISS	www.kriss.re.kr
Sweden	SP	www.sp.se
Switzerland	METAS	www.metas.ch
Taiwan	ITRI	www.itri.org.tw
United Kingdom	NPL	www.npl.co.uk
	NEL	www.tuvnel.com
USA	NIST	www.nist.gov

Note: More links and updated information can be found at *www.measurement.academy*.

Glossary

Actuator

An actuator is a component that is responsible for moving a mechanism or component part – a sort of motor. An actuator on a valve will automatically open or close the valve. It can be hydraulic, pneumatic or electric.

Analogue

An analogue (or analog) signal is a signal without 'steps'. In a way, analogue is the opposite of digital. However, in practice, most analogue signals are digital anyway, since they are produced in a digital converter. With a high-resolution converter, there are many steps and the signal quality is good.

Arbitrary

A method or data agreed upon (or commonly used) but not based on a fixed standard or law.

Autoclave

An autoclave is a pressurised chamber that can be used for high temperature cleaning (sterilisation) of various tools and components.

Batch

Batch means 'dose', 'volume' or 'part' in a manufacturing process in which components or goods are produced in batches (groups) and not in a continuous stream.

By-pass

By-pass basically means a way to go around. In a process industry, it mainly refers to a valve that (if it is open) lets the fluid pass a specific component (such as a filter). In a pump, a by-pass valve will allow some of the flow to flow from outlet back to inlet, and this will result in a possible way to adjust the pressure generated by the pump.

Capacity

The capacity of a tank or container is equal to the total amount of space. When the tank is partly filled, the *volume* refers to the content in the tank (the quantity of the stored product) and the *ullage* is the remaining space.

Coaxial

A coaxial cable is a single conductor wire with a screen around. A coaxial pipe is similar, a rod inside a tube. Coaxial cables are in general used to transfer signals with high frequency.

Coefficient

A coefficient is a multiplicative factor, a number describing the impact of a certain effect. For example, a friction coefficient describes the ratio of the force of friction between two bodies.

Condensation

If air or gas is cooled, there is no room for the same amount of water as in hot air or gas. The water then condenses and forms liquid. Condensate inside a pipe or an instrument must be avoided as it will easily cause problems. As minerals and other parts of ordinary tap water will not follow, condensate has other properties, for example, the conductivity is very low.

Contamination

A contaminated device is approximately equal to a dirty device, but in most cases the contaminated device is destroyed as it is not possible to clean it.

Differential

By using 'differential technique', many measuring problems can be avoided. Let's take a pressure sensor affected by ambient temperature as an example, that is, the measuring error is larger at 30 °C than at 20 °C. If we now install two such sensors, where only one is connected to the process and the other to the atmosphere, we can instead look at the difference between the two pressure readings. By doing this, we now have a value independent of temperature! This assumes that the temperature dependency is repeatable and systematic.

EU

The member states of the European Union (EU) are Austria, Belgium, Bulgaria, Croatia, Cyprus, Czech Republic, Denmark, Estonia, Finland, France, Germany, Greece, Hungary, Ireland, Italy, Latvia, Lithuania, Luxembourg, Malta, the Netherlands, Poland, Portugal, Romania, Slovakia, Slovenia, Spain, Sweden and the United Kingdom (2016).

Fluid

A fluid can be either liquid or gas. In a gas, molecules are far apart from each other which make the gas compressible and it will expand if external pressure is removed. In a liquid, molecules are closer to each other and the liquid will not expand if external pressure is removed. Also a liquid can be compressed, but only slightly. Temperature and pressure will affect these properties and can even transform a gas to liquid or vice versa. A *vapour* is a gas with properties close to liquid.

Gauge

A gauge can not only refer to a scale of measure (most often pressure) but also be a device for measuring pressure, temperature or flow.

Impedance

Internal resistance and capacitance or inductance result in impedance. For example, a loudspeaker in a HIFI system is labelled with an impedance. Also other electrical system (inputs/outputs) has impedance limits specified.

Ionisation

Ionisation is the process that removes electrodes from atoms. The atoms become positively charged and are called ions.

Kinetic Energy

The kinetic energy of an object is the energy that it possesses due to its motion.

Linear

A linear relation is a relation that can be illustrated in a diagram with a straight line. In a measuring instrument, it is good if the input (e.g. temperature) and output (e.g. mA-signal) have a linear relationship.

Magnetite

Magnetite is a mineral and is found in iron ore. However, it is also the name of deposits found on the interior of pipes in a heating system. Since it contains iron, it will conduct electrical current and thereby act like a short circuit in, for example, an inductive flow meter.

Metrologist

A metrologist is a person who works with, or studies, metrology, measurement technology and/or weights and measures. Observe that meteorology is different, as this refers to atmosphere and weather phenomena.

Mock-Up

A scale or a full-size copy of a pipe work (or any design), built with the purpose to test or demonstrate functions and characteristics.

Molecule

A molecule is a group of atoms, having a specific property. There are small groups, such as hydrogen with only two atoms, or large groups, such as petroleum products. It is molecules that make it possible for us to perceive both temperature and pressure.

Momentary

Momentary means lasting for very short time, brief. A momentary signal will always follow the process value, for example, pressure or temperature. A totaliser is different, and it will instead add more to the total sum with an increase in process value. When process value goes to zero, the momentary value will also go to zero, and the totaliser will stop at the present value.

Phase Shift

A phase shift (or difference) can be observed when comparing two signals and the delay time for one of them is getting bigger or smaller.

Pitot

A pitot tube is a kind of flow meter, a pipe with an open end pointing against a fluid in motion. There is also another opening not exposed to the motion. Because of momentum energy, there will be a difference in pressure. A Prandtl tube is a similar device, but with many holes exposed to the fluid in motion, resulting in an averaging effect.

Pneumatic

Energy (or signal) transferred by pressurised air. Many tools are pneumatic and driven by compressed air.

Precise

A precise measure is close to an exact measure. In everyday language, precise is used for devices with small errors, for example, a ruler with small divisions (such as millimetres). The word precious is different, with the meaning valuable.

Pressure Equipment

Pressure equipment means steam boilers, tanks, piping, safety valves and other components subject to pressure loading. All such devices must be produced in a controlled way according to an approved design. Small items and items exposed to a limited pressure are not included.

Pulsation

Pulsation is a periodic increase and decrease in pressure, temperature, volume or similar. In a process, this is mainly (but not always) something unwanted creating problems for both instruments and other components.

Redundant

A redundant system is prepared for malfunction. It can be, for example, two microprocessors, where one is in standby mode, prepared to start up if the master stops working. Redundant systems, with double sensors or actuators, are common in safety applications.

Reference Value

Here a reference value refers to a 'true' value. After reading this book, you will know that hardly anything is absolutely true, so to be precise a reference value is what in 'general is considered to be true'. This is a very fundamental explanation, normally not needed in ordinary work.

Repeatable

If doing the same thing over and over with the same result, the process is repeatable. In measurements, this is very good. If you, for example, put the same weight on a weighing scale many times, of course you want the display to always indicate the same mass. A weighing scale with good repeatability will do this.

RH – Relative Humidity

Moisture in air and other substances is measured in relative humidity. 100% RH for air is equal to the maximum water content without condensation (forming droplets). This varies with temperature, and at 20 °C the maximum water content in air is around 18 g/m^3.

Risk Assessment

A risk assessment is an overview of things that can go wrong and possible consequences. This analysis can be done in several ways, all based on common sense. However, some specially designed models are available that help not to overlook anything.

Seal

Mark or brand made by an inspector of weights and measures to show that the instrument is tested and approved. In programmable devices, the seal can also be in the form of a password.

Standard

The word standard can refer to several things. A document that describes a common method or design is a standard. An established measure that acts as reference (or 'true') when measuring, for example, length, volume and mass is also a standard.

Systematic

Systematic in general terms means in accordance with a fixed procedure or principle. A systematic error has a similar definition and may occur because there is something wrong with the instrument or because the instrument is wrongly used. Other types of errors are unpredictable and random.

Tampering

To tamper with an instrument is not allowed as it means interfere with or to forge the data.

Thermal

Energy transferred by a hot or cold fluid. One example is a steam turbine where hot steam is used to generate mechanical power. Another example is district heating, where homes are heated from a centrally located heat plant.

Time Constant

All instruments have an internal processing time, which results in a delay between an actual change in process conditions and a change in the output signal. This delay can be described and stated in many ways. Response time in a data sheet can be the time between 10% and 90% of signal change for a sudden change of input. It can also be the time from 0% to 100% or 0% to 63% (common for mechanical systems).

Traceability

Traceability is very important for all measuring instruments. Traceable measurements are securely linked to references and international standards. Without traceability measuring, results cannot be compared with measurements performed by other instruments.

Tolerance

A tolerance is an allowed error or an accepted deviation.

Verify

If your job is to verify an instrument, you shall check that it is working and in good condition. Also you shall check if it is measuring correctly; that is, you shall calibrate it and make sure measuring errors are within stated limits. Verification can also include a check of serial number, software version and configuration settings.

Appendix: Physical Data and Material Properties

All tables present indicative, general data only (Tables A.1–A.30). For verified and traceable information, you will need to contact the supplier of the specific product used in your process.

WARNING: Failure to do so may result in serious injury.

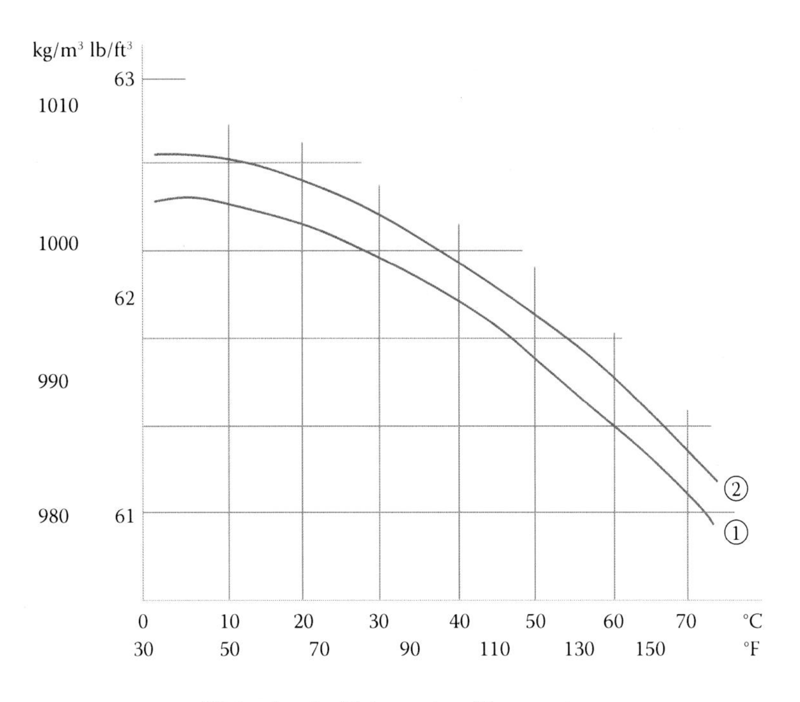

Water density (1) tap water, (2) sea water.

TABLE A.1

Density and Viscosity of Water

Temperature		Density[a]		Density[b]	Viscosity
°C	°F	g/cm^3	pounds/ft^3	g/cm^3	cSt
0,0	32,0	0,99987	62,416	0,99984	
4,0	39,2	1,00000	62,424	0,99976	1,6
4,4	40,0	0,99999	62,423	0,99974	
10,0	50,0	0,99975	62,408	0,99970	1,3
15,6	60,0	0,99907	62,366	0,99901	
21,0	70,0	0,99802	62,300	0,99799	1,0
26,7	80,0	0,99669	62,217	0,99660	
32,1	90,0	0,99510	62,118	0,99500	0,8
37,8	100	0,99318	61,998	0,99305	
48,9	120	0,98870	61,719	0,99854	0,6
60,0	140	0,98338	61,386	0,98321	
71,1	160	0,97729	61,006	0,97715	0,4
82,2	180	0,97056	60,586	0,97042	
93,3	200	0,96333	60,135	0,96307	0,3
100	212	0,95865	59,843	0,95835	

[a] "pure" water (depending on quality and mineral content). Sea water contain salt and have a density which is approximately 2,7% higher.

[b] distilled water (reference data from the International Association for the Properties of Water and Steam).

TABLE A.2

Density and Viscosity of Air at Atmospheric Pressure

Temperature		Density, Dry Air (0% RH)		Humid Air (50% RH)	Humid Air (100% RH)	Viscosity
°C	°F	kg/m^3	pounds/ft^3	g/cm^3	g/cm^3	cSt
−10	14	1,341	0,084	1,340	1,340	12,3
0	32	1,292	0,081	1,291	1,289	13,2
10	50	1,247	0,079	1,244	1,241	14,0
15	59	1,225	0,076	–	–	18,0
20	68	1,204	0,075	1,199	1,193	15,0
30	86	1,164	0,073	–	–	16,1

TABLE A.3

Density and Viscosity of Selected Liquids (at 20 °C)

Substance	Density (kg/m^3)	Viscosity (cSt)
Acetone	790	0,5–1,3
Ethyl alcohol	850	1–2
Ammonia 10%	900	1
Benzaldehyde 0,1%	1050	1
Blood	1060	5–10
Bromine	3100	0,3
Butane	580	0,1
Butanol	810	1,2
Cyclohexanol	960	70
Diesel oil	830	5
Ethanol	790	5
Ether	710	1
Ethylene glycol	1110	95
Phenol	1070	10
Fish oil (liver)	930	35
Kerosene	780	2,5
Glycerin	1260	500–1000
Heptane	680	15
Hexane	660	0,6
Honey	1400–1600	1000–2000
Isobutanol	810	12
Isopropanol	780	3
Peanut oil	910	45
Chloroform	1480	0,4
Coconut oil	930	60
Linseed oil	930	30–50
Mayonnaise	900–1100	5000–6000
Methanol	790	0,8
Milk	1003–1040	2–4
Formic acid	1220	1,5
Sodium hydroxide 20%	1220	0,8
Sodium hydroxide 40%	1430	14
Sodium hydroxide 50%	1530	25
Sodium hydroxide 100%	2130	–
Propylene glycol	1040	55
Nitric acid 10%	1050	5
Nitric acid 100%	1510	–
Hydrochloric acid 20%	1100	1

(Continued)

TABLE A.3 *(CONTINUED)*

Density and Viscosity of Selected Liquids (at 20 °C)

Substance	Density (kg/m^3)	Viscosity (cSt)
Syrup dark @ 20 °C	1300–1400	7800
Syrup dark @ 40 °C	1300–1400	900
Butter oil	860	100000
Sulphuric acid 96%	1840	50
Sulphuric acid 98%	1840	16
Acetic acid	1050	2
Beer	1010	2
Fuel oil (distillate)	850–950	2–15

TABLE A.4

Density, Viscosity and Isentropic Coefficient of Selected Gases (at 0 °C and 0 bar g)

Substance		Density (kg/m^3)	Viscosity (cP)	K[a] (Cp/Cv)
Acethylene	C_2H_2	1,17	0,009	1,2
Ammonia	NH_3	0,77	0,009	1,4
Argon	Ar	1,78	0,021	1,6
Nitrogen	N_2	1,25	0,017	1,4
Butane	C_4H_{10}	2,67	0,008	1,1
Ethane	C_2H_6	1,36	0,008	1,2
Ethylene	C_2H_4	1,26	0,010	1,2
Helium	He	0,18	0,019	1,7
Chlorine	Cl_2	3,22	0,012	1,4
Oxygen	O_2	1,43	0,020	1,4
Methane	CH_4	0,72	0,010	1,3
Ozone	O_3	2,14	0,030	1,7
Propane	C_3H_8	2,02	0,008	1,1
Carbon dioxide	CO_2	1,98	0,014	1,3
Carbon monoxide	CO	1,25	0,017	1,4
Hydrogen	H_2	0,09	0,008	1,4

[a] isentropic coefficient/heat capacity ratio.

TABLE A.5

Electrical Conductivity of Selected Liquids

Substance	Temperature (°C)	Conductivity (μS/cm)
Acetic acid (1%)	18	300
Acetone	18	0,02
Black liquor	90	5000
Ethyl alcohol	25	0,001
Glycol, ethylene	20	1
Petroleum		<0,001
Water, city	25	10–500
Water, dist		<0,04

TABLE A.6

Conductivity of Selected Solid Materials (at 20 °C)

Substance	Conductivity (S/m)
Silver	$6{,}3 \times 10^7$
Copper	$6{,}0 \times 10^7$
Gold	$4{,}1 \times 10^7$
Aluminium	$3{,}5 \times 10^7$
Platinum	$0{,}9 \times 10^7$
Steel	$1{,}4 \times 10^7$
Mercury	$0{,}1 \times 10^7$
Silicon rubber	$1{,}6 \times 10^{-3}$
Rubber	1×10^{-14}
PTFE	1×10^{-24}
Air	5×10^{-15}

TABLE A.7

pH of Selected Liquids

pH	Substance	
3	Lemon juice	Acidic
5	Coffee	
7	Water	Neutral
9	Soap	Basic
11	Household cleaner	

TABLE A.8

Dielectric Constant of Selected Materials

Substance	ε
Acetone	1
Air	1
Aluminium hydroxide	2,2
Ammonia	16–25
Ash	1,5–2,5
Asphalt	2,5–3,5
Butane	1,4
Butanol	18
Calcium	3
Carbon dioxide	1,6
Chlorine	1,5
Ethanol	24
Gasoline	2
Glass	4–14
Hydrochloric acid	4,6
Hydrogen	1–2
Kevlar	3,4–4,5
LPG	1,6–1,9
Methane	1,7
Neoprene	4–7
Olive oil	3
Paper	1–3
Paraffin	2–3
Petroleum	1,5–2,5
Polyethylene	2–3
Quartz	4–5
Rubber	2–7
Sand	4–6
Silicone	3–5
Sulphur dioxide	16
PTFE	2
Water, distilled	30–80
Wheat flour	3–5
Wood (dry)	1–6

TABLE A.9

Speed of Sound at 20–25 °C

Material	v (m/s)	Liquid	v (m/s)	Gas	v (m/s)
Copper	3000–6000	Water	1495	Air −10 °C	325
Steel	5000–6000	Water, sea	1534	Air 0 °C	331
Aluminium	3000–6000	Diesel	1324	Air 10 °C	337
Glass	4000–5000	Ethanol	1207	Air 20 °C	343
Cork	300–500	Gasoline	1171	Air 30 °C	349
Concrete	3000–4000			Helium	965
Gold	3200			Hydrogen	1270
Wood	3000–4000				
Rubber	100–1800				

TABLE A.10

Heat Capacity of Water

Temperature (°C)	Enthalpy (J/kg °C)	Pressure (kPa) (a)
0,01	4218	100
10	4192	100
20	4182	100
30	4178	100
40	4179	100
50	4181	100
100	4216	101
150	4342	476
200	4497	1550

TABLE A.11

Emissivity

Material	ε
Black body	1,00
Human skin	0,98
Water	0,95
Metal, painted black	0,9
Paper, cardboard	0,8
Copper, oxidised	0,7
Stainless steel	0,6
Galvanised steel	0,5–0,3
Stainless steel, polished	0,1

TABLE A.12

Nominal Resistance of a PRT (IEC 60751)

Temperature		Resistance (ohm)		
°C	°F	Pt-100	Pt-500	Pt-1000
−20	−4	92,160	460,8	921,6
−10	14	96,086	480,4	960,9
0	32	100,00	500,0	1000,0
10	50	103,90	519,5	1039,0
20	68	107,79	539,0	1077,9
50	122	119,40	597,0	1194,0
100	212	138,51	692,5	1385,1

TABLE A.13

Precision Classes of Pt-100 Sensors (IEC 60751)

	Max Deviation (°C)				
	Class				
Temperature (°C)	A	B	1/3 B	1/6 B	1/10 B[a]
0	±0,15	±0,3	±0,1	±0,05	±0,03
100	±0,35	±0,8	±0,6	±0,55	±0,53
200	±0,55	±1,3	±1,1	±1,05	±1,03
400	±0,95	±2,3	±2,1	±2,05	±2,03

[a] 1/10 of class B can normally not be guaranteed in a process application.

TABLE A.14

ITS-90 Calibration Points (CIPM)

	Temperature			
Material	**K**	**°C**	**°F**	**Type**
Hydrogen (H)	13,8033	−259,3467	−434,8241	Triple point
Neon (Ne)	24,5561	−248,5939	−415,4690	Triple point
Oxygen (O)	54,3584	−218,7916	−361,8249	Triple point
Argon (Ar)	83,8058	−189,3442	−308,8196	Triple point
Mercury (Hg)	234,3156	−38,8344	−37,9019	Triple point
Water (H_2O)	273,16	0,01	32,02	Triple point
Gallium (Ga)	302,9146	29,7646	85,5763	Melting point
Indium (In)	429,7485	156,5985	313,8773	Freezing point
Tin (Sb)	505,078	231,928	449,470	Freezing point
Zinc (Zn)	692,677	419,527	787,149	Freezing point
Aluminium (Al)	933,473	660,323	1220,581	Freezing point
Silver (Ag)	1234,93	961,78	1763,20	Freezing point
Gold (Au)	1337,33	1064,18	1947,52	Freezing point
Copper (Cu)	1357,77	1084,62	1984,32	Freezing point

TABLE A.15

Thermal Expansion of Selected Materials

Substance	**Length Expansion (%/°C)**	**Volume Expansion (%/°C)**
Steel	0,0011	0,0034
Stainless steel	0,0017	0,0051
Copper	0,0017	0,0050
Glass	0,0007	0,0021
Plastic (PVC)	0,0052	0,0156

TABLE A.16

Properties of Steam

Pressure (Bar g)	Temperature (°C)	Density (kg/m³)	Viscosity (cP)	Isentropic Coefficient	Compressibility
1,0	120	1,14	0,013	1,31	0,98
2,0	134	1,66	0,013	1,31	0,97
3,0	145	2,17	0,014	1,31	0,96
4,0	152	2,67	0,014	1,31	0,96
5,0	159	3,17	0,014	1,30	0,95
6,0	165	3,67	0,015	1,30	0,94
7,0	171	4,17	0,015	1,30	0,94
8,00000	175	4,66	0,015	1,30	0,93
9,00000	180	5,15	0,015	1,30	0,93
10,0000	184	5,64	0,015	1,30	0,92
11,0000	188	6,13	0,015	1,30	0,92
12,0000	192	6,62	0,015	1,30	0,92
13,0000	195	7,11	0,016	1,30	0,91
14,0000	198	7,60	0,016	1,29	0,91
15,0000	201	8,09	0,016	1,29	0,90
16,0000	204	8,58	0,016	1,29	0,90

TABLE A.17

Flow and Velocity

Dimension		Flow Rate				
DIN	ANSI	l/min	m³/h	l/s	UK gallons/h	US gallons/h
10	3/8″	4,71	0,28	0,078	62,2	74,7
15	1/2″	10,6	0,64	0,177	139,9	168,0
25	1″	29,4	1,77	0,491	388,0	466,0
40	1 ½″	75,4	4,52	1,257	995,1	1195
50	2″	118	7,07	1,964	1557	1870
80	3″	302	18,1	5,026	3986	4787
100	4″	471	28,3	7,854	6216	7466

TABLE A.18

Electrical Cables/Typical Cable (Cu) Resistance

Area (mm^2)	AWG[a] Approx.	Resistance (mΩ/m)	Approx. Max Load (A)[b]
0,5	20	35	1,5–3
0,75		22	3–6
1	17	17	5–10
1,5		11	8–16
2,5	13	7	
4		4	

[a] American Wire Gauge.
[b] Observe local regulations.

TABLE A.19

Electrical Cables/Cable Color Codes

Color	Code
Blue	BU
Brown	BN
Green	GN
Yellow	YE
Red	RD
Black	BK

TABLE A.20

Air Buoyancy Effect

Material	Density (kg/m^3)	Correction
Gold	19300	−0,1%
Steel	8000	±0,0%
Water	1000	+0,1%
Wood	500	+0,2%
Styrofoam	30	+4,2%

TABLE A.21

Pipe Dimensions – Outer Diameter

	Outer Diameter			
Nominal Size	**ISO**	**DIN**	**ASTM**	**EN**
1/2″ (DN15)	21,3	20,0	21,3	21,3
3/4″ (DN20)	26,9	25,0	26,7	26,9
1″ (DN25)	33,7	30,0	33,4	33,7
1 ½″ (DN40)	48,3	44,5	48,3	48,3
2″ (DN50)	60,3	57,0	60,3	60,3
3″ (DN80)	88,9	88,9	88,9	88,9
4″ (DN100)	114,3	108,0	114,3	114,3
6″ (DN150)	168,3	159,0	168,3	168,3
8″ (DN200)	219,1	216,0	219,1	219,1
10″ (DN250)	273,0	267,0	273,0	273,0
12″ (DN300)	323,9	318,0	323,9	323,9

TABLE A.22

Pipe Dimensions – Inner Diameter EN[a]

	Class					
Nominal Size	**1**	**2**	**3**	**4**	**5**	**6**
1/2″ (DN15)		17,3	16,1	14,9	13,3	21,3
3/4″ (DN20)		22,3	21,7	20,5	18,9	17,9
1″ (DN25)		28,5	27,3	25,7	24,7	22,5
1 ½″ (DN40)		43,1	41,1	40,3	38,3	35,7
2″ (DN50)		54,5	53,1	52,3	49,1	46,1
3″ (DN80)		82,5	80,9	77,7	72,9	71,3
4″ (DN100)		107,1	105,3	101,7	96,7	92,3
6″ (DN150)	160,3	159,3	157,1	154,1	146,3	139,9
8″ (DN200)	210,1	206,5	204,9	203,1	194,1	187,1
10″ (DN250)	263	260,4	255,4	253	248	241
12″ (DN300)	312,7	309,7	306,3	303,9	298,9	288,9

[a] DIN EN 10253–2.

TABLE A.23

Pipe Dimensions – Inner Diameter ISO[a]

	Class				
Nominal Size	**1**	**2**	**3**	**4**	**5**
1/2″ (DN15)	18,1		17,3	14,9	13,3
3/4″ (DN20)	23,7		22,3	20,5	18,9
1″ (DN25)	29,7		28,5	27,3	25,7
1 ½″ (DN40)	44,3		43,1	40,3	38,3
2″ (DN50)	56,3		54,5	51,5	49,1
3″ (DN80)	84,3		82,5	77,7	72,9
4″ (DN100)	109,1		107,1	101,7	96,7
6″ (DN150)	163,1	160,3	159,3	154,1	146,3
8″ (DN200)	213,3	210,1	206,5	203,1	194,1
10″ (DN250)	267,2	263	260,4	255,4	244,6
12″ (DN300)	318,1	312,7	309,7	303,9	291,9

[a] ISO 1127.

TABLE A.24

Pipe Dimensions – Inner Diameter ASME[a]

	Class					
Nominal Size	**20**	**40/STD**	**80/XS**	**120**	**160**	**/XXS**
1/2″ (DN15)		15,8	13,9		11,8	6,4
3/4″ (DN20)		20,9	18,8		15,5	11,0
1″ (DN25)		26,6	24,3		20,7	15,2
1 ½″ (DN40)		40,9	38,1		34,0	27,9
2″ (DN50)		52,5	49,3		42,9	38,1
3″ (DN80)		77,9	73,7		66,7	58,5
4″ (DN100)		102,3	97,1	92,1	87,5	80,1
6″ (DN150)		154,1	146,3	139,7	131,7	124,3
8″ (DN200)	206,3	202,7	193,7	182,7	173,1	174,7
10″ (DN250)	260,2	254,4	242,8	230,2	215,8	222,3
12″ (DN300)	311,1	303,3	288,9	273,1	257,3	

[a] ASME B 36.10 – 19.

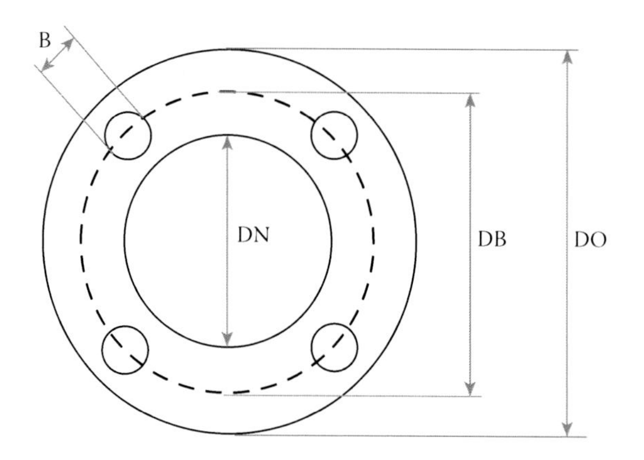

FIGURE A.1
Nominal flange dimensions (Tables A.25–28).

TABLE A.25

Flange Dimensions – DIN, PN6, PN10 and 16[a]

DIN	PN6			PN10			PN16		
DN	DO	DB	BOLTS	DO	DB	BOLTS	DO	DB	BOLTS
10	75	50	4 × 11	90	60	4 × 14	90	60	4 × 14
15	80	55	4 × 11	95	65	4 × 14	95	65	4 × 14
20	90	65	4 × 11	105	75	4 × 14	105	75	4 × 14
25	100	75	4 × 11	115	85	4 × 14	115	85	4 × 14
40	130	100	4 × 14	150	110	4 × 18	150	110	4 × 18
50	140	110	4 × 14	165	125	4 × 18	165	125	4 × 18
80	190	150	4 × 18	200	160	4 × 18	200	160	8 × 18
100	210	170	4 × 18	220	180	8 × 18	220	180	8 × 18
150	265	225	8 × 18	285	240	8 × 22	285	240	8 × 22
200	320	280	8 × 18	340	295	8 × 22	340	295	12 × 22
250	375	335	12 × 18	395	350	12 × 22	405	355	12 × 26
300	440	395	12	445	400	12 × 22	460	410	12 × 26

[a] Nominal dimensions in inches, other dimensions in mm including number and size of bolts in mm. (size, DO, DB – see Figure A.1) a) nominal pressure in bar.

TABLE A.26

Flange Dimensions – DIN, PN25 and PN40[a]

DIN	PN25			PN40		
DN	DO	DB	BOLTS	DO	DB	BOLTS
10	90	60	4 × 14	90	60	4 × 14
15	95	65	4 × 14	95	65	4 × 14
20	105	75	4 × 14	105	75	4 × 14
25	115	85	4 × 14	115	85	4 × 14
40	150	110	4 × 18	150	110	4 × 18
50	165	125	4 × 18	165	125	4 × 18
80	200	160	4 × 18	200	160	8 × 18
100	235	190	8 × 22	235	190	8 × 22
150	300	250	8 × 26	300	250	8 × 26
200	360	310	12 × 26	375	320	12 × 30
250	425	370	12 × 30	450	385	12 × 33
300	485	430	16 × 30	515	450	16 × 33

[a] Nominal dimensions including number and size of bolts in mm. (size, DO, DB – see Figure A.1) a) nominal pressure in bar.

TABLE A.27

Flange Dimensions – ANSI[a]

ANSI	150 lbs			300 lbs			600 lbs		
NPS	DO	DB	BOLTS	DO	DB	BOLTS	DO	DB	BOLTS
1/2″	89	60	4 × 16	95	67	4 × 16	95	67	4 × 16
3/4″	99	70	4 × 16	117	83	4 × 20	117	82	4 × 19
1″	108	80	4 × 16	124	89	4 × 20	124	89	4 × 19
1 ½″	127	98	4 × 16	156	114	4 × 23	156	114	4 × 22
2″	152	121	4 × 20	165	127	8 × 20	165	127	8 × 19
3″	190	152	4 × 20	210	168	8 × 23	210	168	8 × 23
4″	229	190	8 × 20	254	200	8 × 23	273	216	8 × 26
6″	279	241	8 × 23	318	270	12 × 23	355	292	12 × 29
8″	343	298	8 × 23	381	330	12 × 23	419	349	12 × 32
10″	406	362	12 × 26	445	387	16 × 32	508	432	16 × 35
12″	482	431	12 × 26	520	450	16 × 35	558	489	20 × 35

[a] ANSI/ASME B16.5.

Note: Nominal dimensions in inches, other dimensions in mm including number and size of bolts in mm. (size, DO, DB – see Figure A.1) a) nominal pressure in bar.

TABLE A.28

Flange Dimensions – JIS[a]

JIS	10 kgf/cm²			16 kgf/cm²			40 kgf/cm²		
10	90	65	4 × 15	90	65	4 × 15	110	75	4 × 19
15	95	70	4 × 15	95	70	4 × 15	115	80	4 × 19
20	100	75	4 × 15	100	75	4 × 15	120	85	4 × 19
25	125	90	4 × 19	125	90	4 × 19	130	95	4 × 19
40	140	105	4 × 19	140	105	4 × 19	160	120	4 × 23
50	155	120	4 × 19	155	120	8 × 19	165	130	8 × 19
80	185	150	8 × 19	200	160	8 × 23	210	170	8 × 23
100	210	175	8 × 19	225	185	8 × 23	250	205	8 × 25
150	280	240	8 × 23	305	260	12 × 25	355	295	12 × 33
200	330	290	12 × 23	350	305	12 × 25	405	345	12 × 33
250	400	355	12 × 25	430	380	12 × 27	475	410	12 × 33
300	445	400	16 × 25	480	430	16 × 27	540	470	16 × 39

[a] Japanese Industrial Standard B2220.
Note: Nominal dimensions including number and size of bolts in mm. (size, DO, DB – see Figure A.1) a) nominal pressure in kgf/cm².

TABLE A.29

Steel-Type Equivalents

	EN10027	AISI/ ASTM[a]	UNS
Stainless steel	1.4301	304	S30400
Stainless steel	1.4401	316	S31600
Stainless steel	1.4404	316 L	S31603
Stainless steel	1.4571	316 Ti	S32100
Inconel®	2.4816		N06600
Hastelloy®	2.4819		N10276
Steel	1.0460	SA105/C22.8	
Steel	1.5415/16Mo3	A204 Gr.A	

[a] Identical or similar properties.

TABLE A.30

Corrosion/Chemical Resistance of Selected Materials

Fluid		Carbon Steel	Copper	St Steel 304	St Steel 316	Hastelloy C®	Titanium	Rubber	Viton®	PTFE (R)
Acetic Acid 20%	CH_3COOH	D	B	B	A	A	A	B	C	B
Acetic Acid 80%	CH_3COOH	D	B	D	B	A	A	C	C	B
Acetic Acid Vapors		–	B	D	D	A	A	A	A	A
Acetone, 50% water	CH_3COCH_3	–	–	B	B	A	A	D	D	A
Acetylene	CH_2	A	D	A	A	–	–	B	A	A
Aluminum Chloride	$AlCl_3$	A	B	B	B	A	B	A	A	A
Aluminum Chloride 20%	$AlCL_3$		–	D	C	A	B	A	A	A
Aluminum Fluoride	AlF_3	D	D	D	D	B	A	B	A	A
Aluminum Hydroxide	$AlOH_3$	–	D	A	C	B	B	D	A	A
Aluminum Nitrate	$AlNO_3$	D	–	A	A	–	A	A	A	A
Aluminum Sulfate	Al_2SO_4	D	A	B	B	B	A	A	A	A
Ammonia 10%	NH_3	–	D	A	A	A	C	D	D	A
Ammonium Chloride	NH_4F	D	D	C	B	D	B	A	A	A
Ammonium Hydroxide	NH_4OH	D	D	A	A	B	B	D	B	A
Ammonium Sulfate	$(NH_4)_2SO_4$	D	D	B	B	B	A	A	A	A
Ammonium Sulfite	$(NH_4)_2S$	D	D	B	B	–	–	A	D	A
Antifreeze ("glycol")		–	–	B	A	A	–	A	A	B
Arsenic Acid	H_3AsO_4	D	A	A	A	B	B	B	A	A
Asphalt		B	A	B	A	–	–	D	A	A
Barium Carbonate	$BaCO_3$	–	A	B	B	B	A	–	A	A
Barium Chloride	$BaCl_2$	C	B	A	A	B	A	A	A	A

(Continued)

TABLE A.30 *(CONTINUED)*

Corrosion/Chemical Resistance of Selected Materials

Fluid		Carbon Steel	Copper	St Steel 304	St Steel 316	Hastelloy C®	Titanium	Rubber	Viton®	PTFE (R)
Barium Hydroxide	$BaOH_2$	A	–	B	B	B	B	A	A	A
Barium Nitrate	$Ba(NO_3)_2$	C	B	B	B	–	A	–	A	A
Barium Sulfate	$BaSO_4$	A	B	B	B	A	B	A	A	A
Barium Sulfide	BaS	D	D	B	B	–	A	A	A	A
Beer		C	B	A	A	A	B	A	A	A
Benzene	C_6H_6	A	B	B	B	B	A	D	A	A
Bleach		D	–	A	A	A	A	D	A	A
Boric Acid	H_3BO_3	D	B	B	A	A	A	A	A	A
Bromine	Br	D	–	D	D	A	D	D	A	A
Butane	C_4H_{10}	A	C	A	A	A	A	D	A	A
Butanol (Butyl Alcohol)	C_4H_9OH	B	–	A	A	B	B	A	A	A
Butter		–	–	C	A	–	–	D	A	A
Butylene		A	D	A	A	–	–	D	–	A
Calcium Bisulfate	$Ca(HSO4)_2$	D	–	B	A	–	–	A	–	A
Calcium Carbonate	$CaCO_3$	B	–	A	B	B	B	A	A	A
Calcium Chloride (30% in water)	$CaCl_2$	–	B	C	B	A	A	A	A	A
Calcium Chloride (saturated)	$CaCl_2$	–	B	B	B	A	A	A	A	A
Calcium Hydroxide	$CaOH_2$	B	B	B	B	A	A	A	A	A
Calcium Hydroxide 10%	$CaOH_2$	–	B	B	B	B	A	A	A	A
Calcium Nitrate	$Ca(NO_3)_2$	B	–	C	B	–	B	A	A	A
Calcium Oxide	CaO	–	–	A	A	A	A	B	B	A

(Continued)

TABLE A.30 *(CONTINUED)*

Corrosion/Chemical Resistance of Selected Materials

Fluid		Carbon Steel	Copper	St Steel 304	St Steel 316	Hastelloy C®	Titanium	Rubber	Viton®	PTFE (R)
Carbon Dioxide (dry)	CO_2	A	–	A	A	A	A	B	B	A
Carbon Dioxide (wet)	CO_2	C	–	A	A	A	A	B	B	A
Carbon Monoxide	CO	A	–	A	A	B	–	D	A	A
Carbonated Water		–	B	A	A	–	–	–	–	A
Carbonic Acid	H_2CO_3	D	–	A	A	A	B	C	A	A
Chloric Acid	$HClO_3$	D	D	D	C	A	–	C	–	A
Chlorine Water	$Cl_2 + H_2O$	D	D	C	C	A	A	C	B	A
Chromic Acid 10%	H_8CrO_5	D	D	B	B	A	B	D	B	A
Chromic Acid 30%	H_8CrO_5	D	D	B	B	D	A	D	A	A
Citric Acid	$C_6H_8O_7$	D	D	B	A	A	A	A	A	A
Coffee		–	–	A	A	A	A	A	A	A
Copper Chloride	$CuCl_2$	–	–	D	D	–	D	C	A	A
Diesel Fuel		A	A	A	A	B	B	D	A	A
Ethane	C_2H_6	–	A	A	A	–	–	D	A	A
Ethanol	C_2H_6O	B	A	A	A	A	A	A	B	A
Ethylene Glycol	$C_2H_6O_2$	B	A	B	B	B	A	A	A	A
Fatty Acids		C	D	B	A	A	B	C	A	A
Ferric Chloride	$FeCl_2$	D	D	D	D	B	A	A	A	A
Fluorine	F	D	C	C	A	B	D	C	C	D
Formaldehyde 100%	CH_2O	D	A	C	A	A	A	C	D	A
Formaldehyde 40%	CH_2O	D	B	A	A	B	B	B	A	A

(Continued)

TABLE A.30 ***(CONTINUED)***

Corrosion/Chemical Resistance of Selected Materials

Fluid		Carbon Steel	Copper	St Steel 304	St Steel 316	Hastelloy C®	Titanium	Rubber	Viton®	PTFE (R)
Fruit Juice		–	A	A	A	A	A	D	A	A
Fuel Oils		A	A	A	A	A	A	D	A	B
Gasoline (high–aromatic)		B	B	A	A	A	B	D	A	B
Gasoline, leaded, ref.		B	B	A	A	A	A	D	A	A
Gasoline, unleaded		B	B	A	A	A	A	D	A	A
Gelatin		D	A	A	A	A	A	A	A	A
Glycerin		A	A	A	A	A	A	A	A	A
Grease		–	A	A	A	A	–	D	A	A
Heptane	C_6H_{12}	A	A	A	A	A	A	D	A	A
Hexane	C_6H_{14}	A	A	A	A	A	A	D	A	A
Hydraulic Oil (petroleum)		A	A	A	A	A	–	D	A	A
Hydraulic Oil (Synthetic)		A	A		A	A	–	D	A	A
Hydrazine	$(NH_2)_2$	D	A	A	A	–	–	C	B	A
Hydrobromic Acid 20%	HCl	D	D	D	D	A	A	A	A	A
Hydrochloric Acid 20%	HF	D	D	D	D	A	D	B	A	B
Hydrofluoric Acid 20%	HFl	D	B	D	D	B	D	C	A	A
Hydrogen Gas	H_2	A	A	A	A	A	A	B	A	A
Hydrogen Peroxide 10%	H_2O_2	D	D	B	B	A	A	B	A	A
Ink		–	A	C	C	–	–	D	A	A
Jet Fuel		A	A	A	A	A	A	D	A	A
Kerosene		A	B	A	A	B	A	D	A	A

(Continued)

TABLE A.30 ***(CONTINUED)***

Corrosion/Chemical Resistance of Selected Materials

Fluid		Carbon Steel	Copper	St Steel 304	St Steel 316	Hastelloy C®	Titanium	Rubber	Viton®	PTFE (R)
Lacquer Thinners		A	A	A	A	A	C	D	D	A
Calcium Hydroxide	$CaOH_2$	D	–	B	B	A	A	B	B	A
Potassium Hydroxide	KOH	D	B	B	A	B	D	B	B	A
Sodium Hydroxide	NaOH	D	B	B	B	C	B	A	B	A
Melamine		–	–		D	–	–	–	A	A
Mercury	Hg	C	D	A	A	A	A	A	A	A
Methane	CH_4	D	–	A	A	A	–	D	A	A
Methanol (Methyl Alcohol)	CH_3OH	A	B	A	A	A	X	B	C	B
Milk		D	D	A	A	A	A	A	A	A
Molasses		B	A	A	A	A	A	A	A	A
Motor Oil		A	–	A	A	–	A	–	–	A
Naphtha		B	A	A	A	B	B	D	A	B
Natural Gas		A	–	A	A	B	–	–	A	A
Nitric Acid (5 to10%)	HNO_3	D	D	A	A	A	A	D	B	B
Nitric Acid (Concentrated)	HNO_3	D	D	A	A	B	A	D	B	B
Crude oil		–	B	D	A	A	A	D	A	A
Diesel oil		A	–	A	A	B	B	D	A	A
Fuel oil		A	A	A	A	A	B	D	B	A
Mineral oil		B	B	A	A	A	A	D	A	A
Olive oil		–	–	A	A	A	A	D	A	A
Palm oil		–	A	A	A	–	A	–	A	A

(Continued)

TABLE A.30 *(CONTINUED)*

Corrosion/Chemical Resistance of Selected Materials

Fluid		Carbon Steel	Copper	St Steel 304	St Steel 316	Hastelloy C®	Titanium	Rubber	Viton®	PTFE (R)
Pine oil		–	–	A	A	–	A	D	A	A
Silicone oil		A	A	A	A	A	–	D	A	A
Transformer oil		–	A		A	B	–	D	A	A
Oxygen	O_2			A	A	A	X			
Ozone	O_3	C	A	B	A	–	–	D	A	A
Paraffin		A	B	A	A	B	A	B	B	A
Pentane	C_5H_{12}	C	–	A	A	A	–	D	A	A
Perchloric Acid	$HClO_4$	D	D	C	C	B	D	–	A	A
Petroleum		C	B	A	A	A	A	D	A	A
Phosphoric Acid (<40%)	H_3PO_4	D	D		C	A	C	B	A	B
Phosphoric Acid (>40%)	H_3PO_4	D	D	B	D	A	C	B	A	B
Potassium Chlorate	$KClO_3$	D	B	B	B	B	A	–	A	A
Potassium Chloride	KCl	D	B	B	A	A	A	A	A	A
Potassium Hydroxide	KOH	C	B	B		B	D	B	B	A
Propane (liquefied)	C_3H_8	B	A	A	A	A	–	D	A	A
Propylene Glycol	$C_3H_8O_2$	B	A	B	B	B	A	A	A	A
Salt Brine (saturated)	$NaCl$	D	B	B	A	A	A	B	A	A
Sea Water		D	B	C	C	A	A	A	A	A
Silicone		A	A	A	A	–	–	C	A	A
Soap Solutions		A	A	A	A	A	A	B	A	A
Sodium Carbonate	Na_2CO_3	B	A	A	A	A	A	A	A	B

(Continued)

TABLE A.30 *(CONTINUED)*

Corrosion/Chemical Resistance of Selected Materials

Fluid		Carbon Steel	Copper	St Steel 304	St Steel 316	Hastelloy C®	Titanium	Rubber	Viton®	PTFE (R)
Sodium Hydroxide (20%)	NaOH	D	A	B	B	B	A	A	C	B
Sodium Hydroxide (80%)	NaOH	D	D	C	B	C	D	A	D	B
Sodium Hypochlorite	NaOCl	D	–	D	D	B	C	C	A	A
Sugar (Liquids)		A	A	B	A	A	–	A	A	B
Sulfur Chloride	SCl_2	D	B	D	D	A	D	D	A	A
Sulfur Dioxide	SO_2	D	B	D	A	C	A	C	A	A
Sulfuric Acid (<10%)	H_2SO_4	D	–	D	B	B	D	A	A	B
Sulfuric Acid (75–100%)	H_2SO_4	D	D	D	D	B	D	D	A	A
Trichloroethylene	C_2HCl_3	B	A	B	B	A	A	D	A	A
Turpentine		B	B	A	A	B	B	D	A	A
Urea	$(NH_2)_2CO$	B	–	B	B	B	A	D	A	B
Uric Acid		–	A	B	B	B	A	D	–	B
Vegetable Juice		B	A	A	A	–	–	–	A	A
Vinegar		C	B	A	A	A	–	B	A	A
Vinyl Chloride	C_2H_3Cl	A	B	B	A	A	A	C	A	A
Water, Deionised	H_2O	A	B	A	A	A	A	A	A	A
Water, Distilled	H_2O	D	B	A	A	A	A	A	A	A
Water, Fresh	H_2O	D	B	A	A	A	A	A	A	A
Water, Salt	H_2O	D	B	B	B	A	A	A	A	A

Note: A = good, B–C = further investigation needed, D = not recommended Always confirm resistance information with your material supplier!

Index